PLANTS PLUS

To Dad,
Happy Christmas 87 from Johnnie.

PLANTS
✳ PLUS ✳

George Seddon and Andrew Bicknell

Collins
8 Grafton Street, London W1X 3LA

Plants Plus
was conceived, edited and designed by
Grub Street, Golden House,
28–31 Great Pulteney Street,
London W1R 3DD

Collins Sons & Co Ltd
London ● Glasgow ● Sydney
Auckland ● Toronto ● Johannesburg

First published in Great Britain 1987
© Grub Street London
© Text George Seddon &
Andrew Bicknell

Art Director: Roger Hammond
Project Editor: Brenda Clarke
Designers: Graeme Andrew
and Chris Lowe
Photographs: Simon Battensby
Illustrations: Claire Wright,
Pam Corfield , Julie Carpenter
and Lynette Conway

British Library Cataloguing in
Publication Data
Seddon, George
    Plants plus.
    1. Plant propagation
    I. Title   II. Bicknell, Andrew
    631.5′3   SB119

    ISBN 0-00-412240-2
    ISBN 0-00-412239-9 Pbk

Typeset by Witwell Ltd, Liverpool

Printed and bound in Great Britain
by Blantyre, Glasgow

Photograph overleaf
Leaf cuttings of *Ficus elastica* rolled and
tied for stability while rooting
(see page 119).

Introduction

Propagation is one of the most enthralling aspects of gardening; but it has the misfortune to be burdened with an uninspiring name. No one has written an epic poem in praise of propagation (if indeed there is any poem on the subject at all). Yet the ways in which plants can reproduce themselves are as mysterious as the birth of a new human being, about which writers have been lyrical for centuries.

Recent emphasis in advisory books on childbirth and child rearing has been with the more practical subjects of ante-natal and post-natal care. Similarly, while *Plants Plus* covers the field of what is officially termed 'propagation', it may instead best be described as a horticultural baby care book.

So far, humans reproduce themselves naturally in only one way — seed, embryo, birth. Plants — in addition to 'sexual propagation' (from seed) — can amazingly multiply themselves by 'vegetative propagation', that is, from little bits of themselves. Some can even do it from single cells, but that needs a lot of high tech human help. The aim of this book is to look at these various forms of propagation and explain how to carry them out with as fair a chance of success as any gardener can expect.

But why propagate? The basic reasons are that it can increase your pleasure in gardening and lower the cost. First consider the enjoyment.

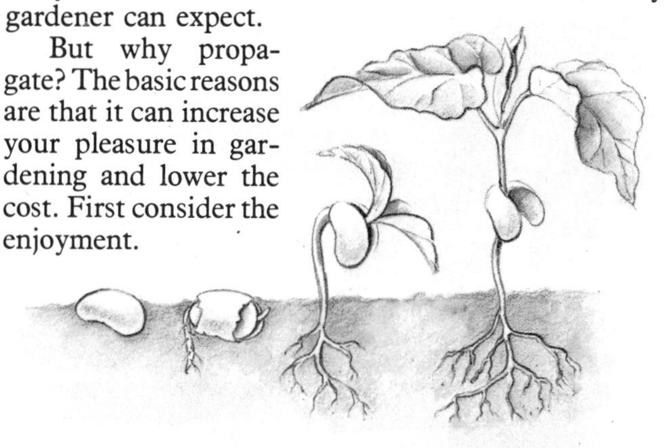

HAPPY NURTURING

If you are lucky, satisfaction as a gardener can start in childhood with the experience of sowing seeds and watching them grow — even humble mustard and cress on a piece of wet cloth in a saucer. Similarly, for other methods of propagation no enormous skills are needed, just care and patience.

Bear in mind too that you always have allies in the seeds and plants you are handling; their determination to live is even greater than yours to see them doing so. In their natural surroundings the going is tough, the chances of survival are minimal and most seeds succumb. (If they did not there would be little room left on earth for us.) So if with your care a seed germinates or a cutting roots, and eventually grows into a magnificent plant, you should be pleased, but not over-proud, for the plant itself has had something to do with it. On the other hand, you should not be downcast by failures; console yourself with the knowledge that you are usually doing better than Mother Nature would. No gardener can attempt more.

The real challenge in propagation is to adjust to the built-in tempo of the plant. The plant will grow at its own rate; it may take three days for a seed to germinate, or three months or three years, and weeks or months for cuttings to root. That you cannot alter.

PLANT ECONOMY

Now for the money saving. Seeds are cheaper to buy than seedlings and infinitely cheaper than established plants. If you already grow annuals from seed, you have the experience to try sowing border perennials, shrubs and even bulbs and some indoor plants. You will not get the same instant results as in buying the finished product from a garden centre or nursery, but you will save a small fortune and have the satisfaction bonus of watching to plant grow: 'Yes, I grew that myself' you will be able to say, self-satisfiedly. Growing vegetables from seed certainly reduces the cost, with the advantage that you can choose varieties

for their good taste and eat them when they are fresh.

Friends will also help to save you money by supplying material for other forms of propagation, offering surplus bits of divided plants, cuttings, offsets and the like. And you will be able to do the same for friends, but try not to thrust on them what you want to dispose of, rather than what they want.

THE WAY TO BEGIN

Some capital investment is involved in propagation. How much depends in the first place on whether you want to propagate only hardy plants, which require the lowest cost, or half-hardy and indoor plants, for which a heated propagator will be needed. Other equipment obviously includes seed trays and pots, and compost for filling them. But the basics should cost little more than several large potted plants. Advice on such equipment, and on suitable composts, is given in the opening pages of *Plants Plus*, 8–23, which also explain in simple terms many of the ways of creating new plants. Having outlined the main forms of propagation, the book then shows how the gardener goes about tackling the methods in both outdoor and indoor gardening.

A detailed opening survey of the relevant methods of propagation is followed in each section by a list of the most popular and interesting plants on which the methods can be tried. Each entry briefly describes the plant, giving information, where applicable, first on the species as a whole and then on individual varieties. It then outlines the way, or ways, in which it can be propagated.

No method guarantees success: some are absurdly easy, others you may want to try 'just for fun'. Either way, you will find your plants far more fascinating as you learn how to recreate them anew.

Contents

Propagation Techniques

Propagation Materials

OUTDOOR PLANTS

INDOOR PLANTS 106

Propagation Techniques

Seeds and Spores

All flowering plants grow from seeds. This is sexual propagation. The male organs of a flower are stamens, at the end of which are the anthers that produce pollen. The female organs include an ovary. When the ovules inside it are fertilized by pollen (usually from another plant) a seed develops.

Not all seeds or spores grow into plants: creating a seed is one thing; launching it successfully into the world is another. Some plants explode their seeds into the air, but that may not carry them far. Many, perhaps with wings or parachutes, are blown on the wind to waft them further afield. Other seeds are eaten by birds and voided elsewhere.

Humans have also spread plants far across the globe, further than Nature could manage, and in the process have changed them so drastically that their ancestors would not recognize themselves for the same plants in the highly-coloured pictures on seed packets.

Even so, seed distribution remains unreliable and chancy and so, not surprisingly, most plants are profligate in producing seed — they might otherwise now be extinct.

But in spite of Nature's and the gardener's efforts, not all seeds sown will germinate. What is remarkable is that so many do.

Human intervention has also produced various forms of treated seed. F_1 hybrid seeds, for example, are crossed from two pure-bred parents; they do not breed true. Dressed seed is coated with fungicide and perhaps insecticide; pelleted seeds are coated with an inert material to enlarge small seeds and make them easier to sow.

Germination of seeds

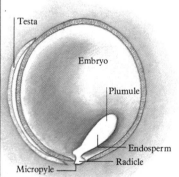

After seeds are sown they need moisture to set off germination. Some absorb water through the outer coat, or testa. Others with harder coats take in moisture through the small pore at the base. Sometimes the coat can be lightly chipped. Germination begins when the radicle breaks through the seed coat and anchors the seed in the soil. Then the plumule pushes up through the soil surface. The radicle always grows down, and the plumule up, whichever way the seed is planted. The radicle then puts out true roots. The first leaves emerge and chlorophyll is produced. This green pigment of plants absorbs energy from the sun for the production of the plant's food. It is a process called photosynthesis.

Cross section of a sweet pea seed, comprising:
Embryo, the rudimentary plant in the seed
Radicle, the embryonic root
Plumule, the embryonic shoot
Endosperm, albuminous storage tissue
Testa, outer covering
Micropyle, pore through which water is absorbed

Dicotyledons

Ferns from spores

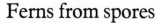

Cotyledons are the first leaves in the embryo of the plant. Some plants have seeds with two cotyledons and are known as dicotyledons, or dicots. They may be herbaceous or woody plants, but their leaves are usually broad with branching veins. Many cotyledons surface above the soil and become the seed leaves. These are often entirely different from the later leaves, the so-called 'true leaves'. After the first true leaves appear the seed leaves die. Not all cotyledons emerge as seed leaves. Some stay underground, feeding the plumule and radicle.

Monocotyledons

Ferns are sexless. They grow from spores, not from seeds, in a most complicated process. Millions of small spores grow in spore cases (sori) on the underside of the fronds. They can be seen as brown spots or patches (above). When ripe, the sori burst, and if the dust-like spores land in a suitably moist place they will in time emerge as prothalli. These look not at all like ferns; their small green leaves (below) resemble clusters of moss. In time, minute male and female sexual organs develop on the underside of the prothalli. The male cells fertilize the female cells, and the resulting offspring create a new generation of ferns — once again sexless.

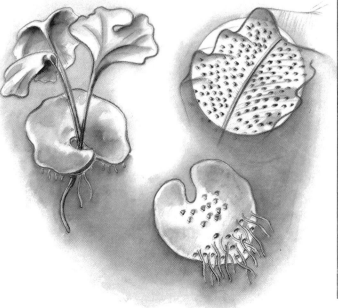

Monocots — properly monocotyledons — have only one seed leaf in the embryo. Among such seeds are grasses and grains, some bulbs and palms, but no true trees. Monocots send one sprout down to become the root and another one up to become the shoot. Its shape does not often differ greatly from later leaves. Unlike the dicots, with their network of veins, monocots have veins running along the length of their leaves.

Another class of plants is the gymnosperms, which includes many conifers. These have not only one or two cotyledons, but even larger numbers.

Collecting and Stratifying Seeds

Collecting your own seeds from plants seems a natural progression from growing your own plants from seeds — and within limits it can be done.

One obstacle may be the climate; in areas where warm, dry summers and autumns cannot be relied on, seeds may not ripen properly. This is particularly the case with late maturing flowers, so the collector may be able to save only the seeds of early maturing plants.

It is also wise to impose this limit on yourself: collect only from **species** plants if you want offspring which look like the parent. Species plants share distinctive features which breed true from generation to generation. What you will get from **varieties** of the species and from **hybrids** is unpredictable. They might be worthwhile growing or, more probably, they could be a dead loss. To duplicate plants which are varieties or hybrids you must use vegetative propagation, which involves taking cuttings, or whatever

other method is suitable. Above all, avoid collecting the seeds of F_1 hybrids, for the results of sowing will be nothing like the parent plant. All this is not to argue against seed collecting but only to point out potential disappointments, some of which can be avoided.

To succeed, first catch your seed. For many plants this must be done in the short critical interval — perhaps only a few days, depending on the weather — between the seed being almost ripe and being shed on to the ground or exploded into the air and lost.

Watch for the signs that indicate whether the seeds are ripening. Most seed-containing capsules — such as pods

— begin to lose their colour, turn brown, shrivel and dry. Others, those of the viola tribe among them, are more difficult to spot because they remain green even when ripe. More frustrating are the exploding capsules which often scatter their seeds without warning, or just as you are gathering them (*impatiens* has this annoying habit). If you have plenty of plants it pays to open a capsule from time to time during the critical period of ripening, to see how the seeds are progressing. As they ripen the seeds also change colour and become drier.

PREPARING SEED

To be on the safe side, especially if the weather threatens to turn wet, the seed heads can be gathered a little before the seeds are fully ripe and taken indoors. Either hang the seed heads in paper (not plastic) bags or spread them over sheets of newspaper in a dry, warm place — in sunshine if possible — to ripen. Obviously the seed heads of different species must never be mixed, but kept in separate bags and immediately and clearly labelled.

The ripe seeds have to be cleaned. While drying, some capsules will have burst open and released their seeds. The rest can be broken up by rubbing them between your hands. Do this over a deep bowl or box to catch any seeds which might explode and carry a fair distance when released.

What you now have are seeds mixed up with a lot of rubbish. Pick out the large pieces of chaff; the smaller stuff can be winnowed by gentle blowing or fanning. Spread the seed and chaff in a shallow tray and slowly move it from side to side, gently blowing the chaff away. 'Gently' is the important word: if you huff and puff or fan furiously you may well blow away the seeds as well.

When that tricky operation is over, put the seeds in paper packets, label them and seal with adhesive tape. Store in a cool dry place, free from frost and mice, until sowing time. Seeds may even be stored in the refrigerator (the top, least-cold shelf is the best), but do not put them in the freezer.

BERRIES AND FRUITS

Extracting seeds from berries and fleshy fruits is more troublesome. They can be pulped with a block of wood (without actually crushing the seeds) put in a bowl of water, and kept in the warmth for a few days. Most of the pulp will rise to the surface and can be skimmed off; most of the seeds will fall to the bottom. Put the seeds in a fine sieve and wash under running water. Spread them on a clean cloth in the warmth to dry, and do this thoroughly or they will go mouldy.

Freshly collected seeds could be sown straight away, and might germinate readily. But there are problems. The seeds of plants in temperate climates ripen at the end of summer. If germinated immediately, they would emerge as seedlings towards the onset of winter, a likely death sentence. Even if grown indoors in heat they would suffer from inadequate light in the winter months. This is why seeds in temperate climates become dormant: they are waiting for warmer days ahead. The seeds themselves can survive hard winters, but the young plants would not.

DORMANCY

A seed will not germinate while it is dormant; that is, while the development of the seed embryo is inhibited. Some seeds become dormant because, while maturing, the seed coat grows thicker and harder, keeping water from the embryo and preventing germination. Another form of dormancy is induced by the manufacture of chemicals within the seed which delay the development of the embryo. Whatever the cause, dormancy has to be broken before the seeds will germinate.

To break the dormancy of seeds with hard seed coats, the outer skin must be scarified to let moisture in. Large seeds — such as those of sweet peas — can be chipped before sowing. A small part of the skin is removed with a file or sharp knife without damaging the embryo. Small seeds can be rubbed over a sheet of sandpaper. The result of both operations is to allow water to reach the embryo.

Given warmth and water it can now grow.

Chemical dormancy is broken by long exposure to cold, by a method known as stratification. This matches the sort of exposure that a seed would experience if it were left lying on the ground after being shed. Stratification also helps to soften hard seed coats.

After extracting the seeds from their capsules, mix them with coarse sand, put them in plastic pots and cover with 1in/2.5cm of sand. Label the seeds in each pot (using indelible ink, for the pots have to stand outdoors) and put them in the coldest spot in the garden, preferably north-facing. Cover securely with small mesh wire netting to keep out birds and, above all, mice.

STRATIFICATION

Alternatively, the seeds of berries and fleshy fruits can be stratified without being extracted. In small plastic pots, put layers of unpulped berries between 1in/2.5cm layers of a moistened mixture comprising one part sand and one part peat. Start and finish with a layer of sand/peat, label and keep outdoors in the cold, protecting the contents from birds and mice.

How long this chilling process must last varies from species to species, and research into this area is patchy. Most seeds need six months of stratification — about the normal length of time between the seed being produced and being sown. Other seeds, among them many shrubs, will need 18 months; in practice that means they have two cold winter periods with a warm summer in between.

When they have been chilled long enough, some seeds start to germinate in the sand/peat in which they have spent the winter. Keep a watch on the pots to see whether any germination is taking place and transfer the contents to seed compost as soon as possible after the first shoots appear. There is no need to separate them from the sand in which they have been stored.

If the seeds show no signs of germination they can be left until you want to sow them, whenever the weather is warm enough.

Division

As well as producing seed, plants have other forms of propagation in their repertoire. These diverse and sometimes surprising methods are lumped together under the description of 'vegetative propagation'. Not all depend on human intervention, although this can be a great help. For example, clumps of plants may spread but, unlike seeds blown by the wind, cannot colonise other parts of the garden. For that they need a gardener, for whom replanting divided clumps elsewhere is the easiest form of propagation.

TYPES OF DIVISION

Other methods of division might be less straightforward, but even so, plants are lavish in offering up pieces of themselves for propagation. They produce **suckers** (shoots growing from the roots); **offsets** (shoots from near the base of a stem); **tubers** and **rhizomes** which can be cut into pieces; small **bulblets** and **cormlets** which grow alongside the parent plants in the soil; and **scales** (the food-storing leaves of lily bulbs) which if removed and planted will also grow bulblets. Even more curiously there are **bulbils**, minute bulbs which grow on the stems, not the bulbs, of a few lilies. Riches, indeed, and almost all obtained from the part of a plant visible to the gardener only when dividing it.

Among plants grown in the herbaceous border, all those which grow several crowns, or clumps, and have fibrous roots can be divided. Those with only one crown, or thick fleshy roots or a tap root, cannot. Lupins (*Lupinus*), carnations, pinks (*Dianthus*) and Oriental poppies (*Papaver orientale*) are therefore excluded from division. Other plants react against root disturbance and may take time to recover; paeonies are an example. The best time to divide most herbaceous perennials is in early spring or spring, when they are just stirring into life.

Many houseplants grow crowns and may be divided at intervals of a few years to save them being moved to ever larger pots. Offsets are the usual way of propagating bromeliads such as *Billbergia*, *Guzmania*, and *Tillandsia*. Orchids, including *Cattleya*, *Cymbidium* and *Dendrobium*, are generally increased by division of rhizomes (pseudobulbs) and so are ferns. Bulbs and corms are prolific in propagating themselves and tubers, with help, will do the same.

Root division

The simplest root division is of fibrous rooted plants, among them many herbaceous perennials. Division is not only propagation, but also rejuvenation. The roots of herbaceous perennials spread year after year, creating an outer rim of young roots and an ageing centre. A plant is divided into the young growth, which is replanted, and the worn-out growth, which is thrown away. The young roots are either replanted as one clump or split into several pieces.

Offsets and suckers

Many plants obligingly grow offsets alongside themselves which can easily be detached and planted to grow into a new plant, a replica of the parent. Bromeliads are notable examples, because some flower, produce offsets and then die. If left to themselves, the general pattern is for parent and offset to live together and spread until parts die of old age. Offsets are therefore an excellent source of new plants. Those from bulbs and corms are shown opposite. **Suckers** are another form of offset and raspberries are a good example. But beware of the suckers of grafted plants, such as many roses. These will be suckers from the rootstock, and not from the grafted growth which you want to reproduce.

Tubers

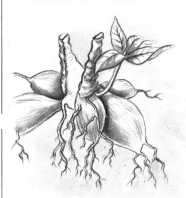

Plants growing from tuberous roots (which are swollen roots or underground stems) can be propagated by dividing the tubers. The basic method is to cut through the crown of the plant so that each section has at least one bud and a piece of tuber. Plants most commonly propagated in this way are dahlias and tuberous begonias.

Rhizomes

Rhizomes are swollen underground stems which grow horizontally. In effect, they reproduce themselves by spreading and branching. To make new plants elsewhere in the garden, lift the rhizomes and cut them into pieces. Each section must be a good size (say 2–3in/5–7.5cm) with at least one strong shoot and preferably more.

Bulbs

Bulbs consist of layers of scales round an embryo stem, from which the flower grows. They propagate themselves naturally by producing one or two offsets. In time, these separate themselves from the parent plant to grow and flower independently. But they can be removed when still small and replanted to grow on their own. They will take from one year to three years or more before flowering. Some bulbs wither after producing flowers, leaving both a new flowering bulb and small ones, which can be removed.

Corms

Corms are swollen underground stems which store food. After flowering, the corms wither. But from the base of each flowering stem, they grow one or two new corms to flower the following year. As well as these instant replacements, many cormlets — very small corms — may also appear between the disintegrating base of the old corm and the new one above it. These are removed, stored in a cold (but frost-free) place in peat for the winter and planted outdoors in spring. They will take two years to flower.

Bulb scales

Lilies, like other bulbs, are made up of scales, which are the fleshy bases of the bulb's

Bulbils

Bulbils are minute bulbs found growing on stems. They are produced by some lilies, including *Lilium tigrinum*, Tiger Lily. Bulbils may occur naturally, or be encouraged by cutting the flowering stems by half before the lily flowers. The bulbils are picked when mature in late summer, stored in peat over winter in the cold and planted outdoors the following spring. They take three years to flower.

leaves. These can be used to grow bulblets, which is like division or taking cuttings. A few scales are removed right down to the base of the bulb and planted, right way up, to half their depth. In time, given warmth, they will root and form a tiny bulb at the base. This is transplanted, hardened off and grown on outdoors, reaching flowering size in about three years.

Cuttings

The idea of new roots growing from old roots seems more likely than new roots growing from shoots. But is it? Plants cannot take stem cuttings of themselves, but if by accident a shoot breaks off and falls to the ground there is a chance, in a suitable environment, that it will put down roots. Such roots are called adventitious — a word used to describe any part of a plant or organ of an animal arising in an unusual place. They are not second class roots, but just as efficient as any other root.

TYPES OF STEM CUTTINGS

The main groups of stem cuttings are **basal** (the baby of the family), **softwood** (juvenile), **semi-ripe** (late adolescent) and **hardwood** (fairly grown up). All these are taken from the current year's growth of the plant, but at different times of the year. Herbaceous perennials provide some basal and many softwood cuttings. Basal cuttings may have been forced early in the year and removed when very small. Softwood cuttings come later, with the soft rapid growth of spring. They are not easy to root, because they lose water through their leaves and quickly wilt, but if too many leaves are removed they are deprived of their source of food. Semi-ripe cuttings usually do better. They are taken from shrubby plants in late summer, when the lower part of the stem is woody and loses less moisture, although the tip is still soft. Hardwood cuttings of deciduous shrubs will be mature, leafless and dormant when they are taken late in the year. At this stage of the season there is no danger from loss of moisture, but rooting will not take place until early the following year.

ROOTING IN WATER

Cuttings, especially of indoor plants, need not be rooted in compost; many will grow roots in water if their lower leaves are removed. Although this may be fascinating to watch, the method has one big drawback. Roots need oxygen. Cuttings in water develop thick white roots which enable them to absorb oxygen from the water. When the cuttings are moved into compost, they must start growing normal roots to absorb oxygen from the air in the compost. There is then an inevitable setback.

Softwood cuttings

Softwood cuttings are young cuttings, taken early in the year from the new growth of the plant, before the stems begin to turn woody. They are usually about 3in/7.5cm long. To root they need not only warmth but high humidity, for if the air is dry the cutting is liable to wilt and die before roots have grown.

Semi-ripe cuttings

Semi-ripe cuttings, usually 4–6in/10–15cm long, are taken later in the year than softwood cuttings. They still come from the current year's growth but the lower part of the stem is now woody, while the tip remains soft. The base of the shoot is cut off just below a node (leaf bud). The soft tip is also often removed, and that cut is made just above a node. This nodal cutting is the type most generally used. Some plants seem to root better from internodal cuttings, however. Here the stem is cut halfway between two nodes. The rate of success from semi-ripe cuttings is generally greater than with softwood cuttings, and because they are taken when the weather is warm, many will root without artificial heat.

Hardwood cuttings

Hardwood cuttings, from 6in/15cm long, are taken late in the year from thoroughly ripe wood of the current year's growth. Those from deciduous plants will probably root better if cut when the leaves are falling or have just fallen. Some can be planted outdoors and will start to grow roots early the next year. Others do better with the protection of a cold frame. A few need heat. By and large, success rate of hardwood cuttings is good.

Heel cuttings

When the side shoots of plants are used as semi-ripe or hardwood cuttings, they are usually taken with a sliver of the bark from the parent stem. With many shrubs especially, this 'heel' encourages the cutting to root. The heel is detached with the shoot by gently pulling it downwards from the main stem, breaking it away as cleanly as possible. Before planting, cut away any rough edges with a sharp knife; cuttings with jagged edges are liable to rot rather than root. To stimulate root growth, the heel cutting is dipped in hormone rooting powder, as recommended for other stem cuttings. These powders are available in strengths appropriate for softwood, semi-ripe and hardwood cuttings.

Mallet cuttings

A semi-ripe cutting with a difference is a mallet cutting — so called from its shape. Taken from the current year's semi-ripe growth, it includes a short section of the woody stem from which it was growing. Thin stems are more reluctant to root than thick ones, and the mallet 'head' makes rooting more likely.

Basal cuttings

Basal cuttings are the small tender shoots which appear from the crown of a plant when growth starts early in the year. When 2in/5cm or so in length, break or cut from the base, preferably with a heel. Half-hardy perennials can be forced in heat from winter onwards, and the shoots removed.

Root cuttings

Many alpines, herbaceous perennials and shrubs can be propagated from cuttings of roots after the plant becomes dormant. Lift small plants from the ground, cut off one or two thick roots, and then replant straight away. If the plant is big, scrape away the soil and remove a few of the outer, younger roots. Cut the roots into 2–3in/5–7.5cm lengths and plant vertically. Cuttings of hardy plants can be planted in a cold frame; the half-hardy need a little heat. Transplant in autumn.

Not only are flowers (or their seeds), roots and stems used for propagation, but some leaves as well. Although the methods are various, there are comparatively few plants which can be reproduced in this way. Most are house or conservatory plants, which require warmth and humidity for success.

VARIETIES OF LEAF CUTTINGS

Some plants are taken **with stalks**; saintpaulias, the African Violets, are an almost cliché example. Roots grow into the compost from the end of the stalk, the leaf being the source of food until the roots are formed and a new plantlet emerges at the base of the leaf. Plantlets are also induced to grow on the leaves of plants with prominent veins, by slashing them. A plantlet can form at the point of each slit. *Begonia rex* is the prime example of this limited form of leaf propagation. One begonia leaf may even be cut into small squares, each capable of producing a plantlet, but more often it happens that the leaf pieces just rot. Leaves of monocot plants can also be cut into small **sections**, each producing a plantlet, but in practice this works only with those few monocots whose thick leaves do not dry out quickly. It is most likely to succeed with the almost indestructible sansevieria, although plantlets from variegated forms will grow just plain green.

Some plants propagated by leaves do not produce plantlets, but these cuttings include not only a leaf but also a leaf bud (from which the new plant grows) as well as part of the parent stem. **Leaf bud cuttings** can propagate a number of houseplants, among the best known being *Ficus elastica* and dracaenas. They need warmth and humidity to root. Some hardy outdoor plants can also be propagated by this method, including ivy. But since ivies can be propagated in many far easier ways, there seems little point in choosing the leaf bud cutting method.

A leaf bud with a short piece of hardwood stem is known as an **eye cutting**. Both fruiting and ornamental vines are increased by these.

There are two especially easy types of cutting suitable for a limited number of plants, although one — the so-called **Irishman's cutting** — is basically a form of division. The other is **piping**, which is used for propagating pinks (*Dianthus*). These cuttings are taken by pulling young shoots away at a node on the stem, instead of cutting them. Irishman's cuttings are taken from plants which grow shoots at their base.

Leaf cuttings with stalk

Of the various ways of taking leaf cuttings, the simplest is to use a leaf with a petiole. The petiole is the stalk by which the leaf is attached to the plant. Leaves which are fully developed but not aged are the most promising to use. After the stalk and leaf are cut from the plant, the stalk is trimmed to about 2in/5cm before being inserted into compost. Most of the few plants propagated in this way are houseplants, and these need fairly high temperatures to root, as well as high humidity to prevent the leaves from drying out. If the cuttings root, a single plantlet, or several, will grow at the base of the leaf.

Leaf cuttings without stalk

The leaves of a few plants, such as *Begonia rex* and species related to it, can be induced to produce plantlets by slashing into the veins. An undamaged, fully grown leaf is used, the stalk removed, and the main veins on the underside cut. The leaf, top side up, is then pinned down on the surface of moist compost so that it stays in close contact with it. As for leaf cuttings taken with stalks, fairly high temperatures and high humidity are necessary for plantlets to develop. These appear where the cuts were made but may take a month or more.

Leaf sections

This method of growing plantlets from leaf cuttings is used for monocot plants, which have veins running the length of the leaves. Cut pieces of leaf *may* form plantlets from the base of each piece. But many thin-leaved plants are more likely to wilt and die before rooting. They include bulbs such as snowdrops, scillas and hyacinths. The thicker leaves of sansevieria do not dry out so rapidly. Cross sections of its leaves, 2in/5cm wide, will produce plantlets in about two months.

Leaf bud cuttings

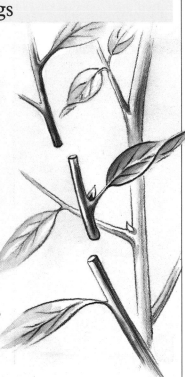

Leaf bud cuttings comprise a leaf, a leaf bud in the axil of the leaf, and a piece of stem just over 1in/2.5cm long. This stem may be softwood, semi-ripe or hardwood. Roots appear from the piece of parent stem and new stems emerge from the leaf bud. The leaf is the source of food until the new roots grow. Therefore the leaf chosen must be mature, or it will use its energy to grow to its proper size. Large leaves can be a nuisance, and are often rolled into a cylinder, bound with a rubber band.
Leaf bud cuttings of such houseplants as *Ficus elastica* and dracaena are taken from spring to summer and will need warmth and humidity to root. Hardy plants — ivy and blackberries especially — are far easier.

Eye cuttings

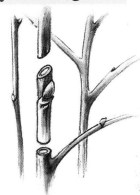

Eye cuttings are short pieces of stem with a bud. Vines, fruiting and ornamental, are propagated this way in early winter or winter. Cut the stem with a bud just below one end and plant vertically in compost. Or cut so that the bud is at the centre of the cutting, and press horizontally into compost.

Pipings

Pipings are the nearest thing to ready-make cuttings, but are provided only by pinks (*Dianthus*). The time to take them is summer and late summer. Pull, not cut, the top of a young shoot with three or four pairs of leaves. It comes clean away at a node. Pipings quickly root in a cold frame and are then planted outdoors.

Irishman's cuttings

This form of simple division can be achieved only with plants which produce shoots from their base or roots. Included among these are Michaelmas daisies and chrysanthemums, among herbaceous perennials, and many alpines. These are the type of shoots which provide basal cuttings, but instead of being cut away to root separately, they are left for a time on the parent plant and then pulled apart from it along with roots to support them. Like other plants gained from simple division, these 'cuttings' can be used to replace ageing clumps.

Layering

Some perennial plants — mainly climbers, shrubs and trees — can propagate themselves by layering. All they need are long trailing or weeping branches which reach the ground. Close contact with the soil produces adventitious roots. The layered branches are only semi-independent, because they are attached to the parent, but as their roots develop they become less dependent. If the old plant dies, the layered young will live perfectly well on its own.

It is not known for how long the phenomenon of such 'simple layering' has been known, but a refined version dates from some 4000 years ago in China. This is air layering, sometimes called Chinese layering. It solved the problem of what to do with a branch which could not be bent down to the ground. Instead it was made to grow roots up in the air. A cut was made in the branch and wet soil packed round it. Roots grew, the branch was cut down, and a new plant was born. Nowadays, adapted to the 20th century, the technique is used on a few houseplants.

With fairly long, bendable stems, simple layering is easy. Longer stems can be buried not just once but at a number of points. This is serpentine layering. The only drawback is that woody stems may be slow to root and, until they do so, will take up a fair amount of room around the parent plant.

Tip layering, in which the tip of a stem is totally buried, is another variant. There are more elaborate versions which are mainly used on a commercial scale. The simplest is called dropping and is used to propagate such low growing shrubs as erica. A plant is dug up in spring and reburied with only the tips of the shoots showing. The shoots root and in autumn are cut away and planted. Stooling involves first growing a cutting, the only purpose of which is to provide a supply of rooted layers year after year. Certain fruit trees and lilacs are among the plants which are propagated in this way.

Outdoor strawberries are natural layers, but some houseplants also produce plantlets at the end of runners or on their leaves. In the wild they might find somewhere to root on their own, but indoors they will need the gardener's help, with pots of compost and bent wire to pin down the plantlets in position on the surface.

Simple layering

Simple layering is simple both in the sense that it produces one more plant, and in being almost absurdly easy. Strawberries produce runners which layer themselves as they spread; brambles put down roots where the tips of their shoots touch the ground. Other plants, with a little help, can be induced to do the same. What is needed is a young shoot near to the ground and soft enough to be bent and buried in the soil. Wait for roots to form before severing the layer from the parent. This may take shrubs from six months to two years.

Serpentine layering

Serpentine layering induces a shoot to root at several points and not just one. The shoot must be both long and pliable, so that it can snake in and out of the several holes made for burying sections of the stem. A slanting cut is made in each section of the stem to encourage rooting. A node buried in the soil produces roots, and for each buried node there must be one or two shoots or buds on the stem which are not buried. It is from these that shoots grow. When well rooted, the layered stem is lifted and split into lengths, with roots and shoots.

Tip layering

Blackberries, and the many hybrid berries bred from them, can be propagated by tip layering. This variation of simple layering is carried out in summer and the whole tip of the shoot is buried. First, the stem is bent down to the ground and at the point where the end reaches the soil, a 4in/10cm deep hole is dug. In this the tip is totally covered with soil; nothing must show above ground. If the shoot has a tendency to jump out of the hole, it can be pinned down. Rooting is rapid and after several weeks a shoot will appear. The rooted layer is now severed from the parent, but then left until autumn before being transplanted.

Air layering

One of the oldest forms of layering has acquired a modern look with the invention of cling film and adhesive tape. Air layering is convenient for plants with branches which cannot be bent down to the ground. However, a common use for it nowadays is to rejuvenate old, overgrown specimens of *Ficus elastica* and dracaenas, especially if they have lost their lower leaves. They are forced to grow new roots higher up the stem. The former top now becomes the new plant and the bottom is usually thrown away. A cut is made in the stem and a ball of damp moss is wrapped round it, held in place by cling film and sticky tape. In time, roots grow from the cut and the now much-shortened plant is potted to start a new life on its own.

Plantlets

Easy layering is provided by the plantlets growing on the stems, runners or leaves of a few houseplants. *Chlorophytum elatum*, Spider Plant (far left) puts out long, flowering stems which grow plantlets at their ends. In time, their weight pulls the stems down and the plantlets can be pinned down in pots of compost. Sever from the parent when rooted. The Pick-a-back Plant, *Tolmiea menziesii* (centre) grows its plantlets at the centre of a leaf. The leaf is pinned down to the surface of a pot of compost, alongside the parent plant, and the runner is cut when rooting has taken place. *Saxifraga stolonifera* (left) grows plantlets on the end of long runners. These can be layered in the same way as *Chlorophytum* plantlets.

Propagation Materials

Compost and pots

Outdoors, gardeners are cursed — or blessed — with the soil they have. They can fertilize or lime it and otherwise try to improve it, but the basic nature of the soil remains little changed, whether clay, chalk, sand or (with luck) loam. What is grown may well be decided more by the gardener's soil than by the gardener.

Indoors, gardeners can easily provide the growing medium best suited to the plants they want to grow. It will not, if they have sense, be soil taken straight from the garden, for there is no point in bringing inside any of the soil-borne diseases which are difficult enough to combat in the garden. Home sterilizing of soil is not foolproof, and in any case, soil in the restricted space of a pot may behave differently from outdoors. It can become compacted with watering, for example, and will not then drain properly, blocking air from the roots and so starving them of oxygen.

It is for this reason that composts have been devised, using a balance of ingredients between those which will not dry out rapidly in a pot and those which will ensure free draining. The other vital ingredient is a balanced fertilizer to meet the food needs of the plant.

As there is such a great choice available of ready-prepared composts, it is probably not worthwhile mixing your own. All the same, it is as well to know what you are buying.

Containers for compost are generally of plastic or clay. Plastic trays and pots are better for seeds and cuttings as they are cheaper, lighter and easier to clean. If plastic is anathema to you, move plants into clay pots later. Peat pots can be planted straight into the garden.

Seed trays and pots

Small plastic pots (say 3in/7.5cm in diameter) are suitable for small sowings of seed or for seedlings. Square pots hold more compost than comparable round pots and take up less space in the propagator or frame. For a larger number of seeds, oblong seed trays are better. Rows of different seeds can be sown on the same tray, but all should be well labelled. Cuttings need deeper compost than do seeds; since most pots are as deep as they are wide, their pots will need to be larger. Several cuttings can share the same pot at the start, however; they are generally spaced round the pot near the rim. Trays with transparent plastic lids act as unheated propagators, providing a slightly warmer and more humid atmosphere.

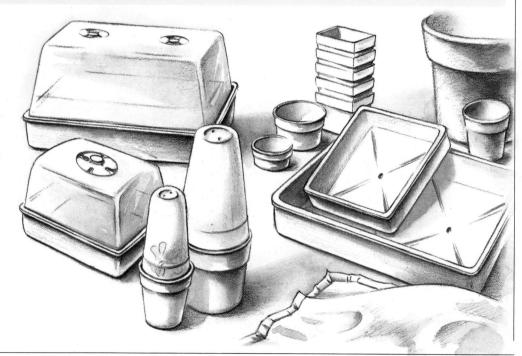

COMPOST MIXES

Water-holding ingredients
Loam Rich soil with plenty of decaying organic material.
Peat Dried, partly decayed remains of sedges and mosses dug up from bogs. Known as peat moss in the United States.
Sphagnum moss Bog plants with leaves that can hold much water. Once a main ingredient of orchid composts.
Sphagnum moss peat Peat from the decayed remains of sphagnum moss.
Tree bark Fine or coarsely ground bark of various trees, now often used in orchid composts.
Vermiculite Flakes of heat-expanded mica which retain large quantities of water.

Free-draining ingredients
Sand Coarse or sharp sand, preferably from lime-free rivers and not seashore sand.

Disposable pots

Plastic and clay pots must be scrubbed clean to avoid disease. Disposable pots cut out this chore. Polystyrene blocks comprise single pots from which the plants have to be removed. Others are made of peat, either pots to be filled with compost, or blocks in which seeds are sown or cuttings inserted. The pot or block is later planted into the soil, hardly disturbing the roots.

Perlite Thermally processed volcanic rock which opens the texture of the soil as well as absorbing water. **Perlag** is a more absorbent version.

Other ingredients
Charcoal Helps to keep compost 'sweet' by absorbing any excess minerals.
Magnesium limestone Reduces the acidity of composts used for plants which need an alkaline soil. Most houseplants require a slightly acid soil.

SEED COMPOSTS

Loam-based seed compost: John Innes formula
2 parts sterilized loam
1 part moss peat
1 part coarse sand
(all measured by bulk, not weight).
To each bushel (36 litres) is added 1.5oz/40g of 18% superphosphate and 0.75oz/20g of chalk.

(A bushel is a dry or liquid measure equal in Britain to 8 Imperial gallons; therefore a bushel of compost would fill four 2-gallon/9-litre buckets. A US bushel is equal to 64 US pints.)

Peat-based seed compost
1 part moss peat
2 parts silver sand
To each bushel can be added — if the seedlings are not to be pricked out quickly — 0.75oz/20g of superphosphate and 1.5oz/40g of ground limestone.

POTTING COMPOSTS

John Innes loam potting composts
These were devised by the John Innes Horticultural Research Institute in the 1930s. All begin with this basic mixture:
7 parts sterilized loam
3 parts peat
1 part coarse sand
To this is added the John Innes Base Fertilizer:
2 parts hoof and horn meal
2 parts superphosphate
1 part sulphate of potash (potassium sulphate)
This fertilizer base is used in varying quantities in the basic compost to make John Innes potting composts 1; and 2 (twice as much fertilizer base as 1); and 3 (three times as much).

John Innes No 1 is for seedlings moved from seed compost and young plants; John Innes No 2 is for mature plants; and John Innes No 3 for large plants in pots over 8in/20cm, or tubs.

Other fertilizer bases are available, giving comparable proportions of nutrients — 5% nitrogen, 7% phosphoric acid, 10% potash — usually labelled 5:7:10 by the makers. Nitrogen promotes growth of leaves and stems, but too much makes them soft. Phosphorus is for healthy root growth. Potash assists flowering and resistance to disease.

Peat-based potting composts
There are endless proprietary brands, but their basic ingredients are peat and coarse river sand, with added fertilizer. The best are made from sphagnum moss peat rather than the less free-draining sedge peat. Sphagnum moss peat is usually less fine and lighter in colour, whereas sedge peat is almost black when wet. Lime-free versions of many composts are produced for use with lime-hating plants.

SPECIAL COMPOSTS

Some plants need, or grow better in, composts specially devised for them. These are available ready prepared, or enthusiasts can make them for themselves.

For bromeliads
1 part coarse leaf mould
1 part moss peat

For cacti and succulents
1 part loam
1 part moss peat
2 parts coarse sand

For ferns
4 parts sterilized loam
4 parts moss peat
3 parts leaf mould
2 parts coarse sand

For orchids
Bark-based: 10 parts shredded pine bark (medium grade) or redwood bark
5 parts of fine grade pine bark
1½ parts of perlag
¼ part of granulated charcoal

Peat-based: 2 parts sphagnum moss peat
1 part coarse sand
1 part perlite

Tools and Equipment

Green fingers may help a gardener but they cannot replace tools. For the simplest propagation — sowing hardy annuals outdoors — the basic equipment is a spade, fork, rake, hoe, trowel and a piece of string if you want to sow in a straight line. Sow half-hardy annuals indoors and you will also need pots or seed trays, compost, fungicide, a simple unheated propagating case (for humidity), some labels and a pencil. To raise tender plants from seed add a heated propagator. Move on to take cuttings and you must invest in a good sharp knife, possibly secateurs, a wide choice of equipment to encourage the cuttings to root, hormone rooting powder and maybe cloches and cold frames to help young plants face life outdoors.

Nature seems to get along reasonably well without all the paraphernalia that gardeners surround themselves with to propagate far fewer plants. We have greater expectations. We fret or sulk when seeds do not germinate or cuttings die, whereas Nature expects millions of seeds to perish without trace. The equipment we use for propagation is meant to help us do better than that, making life easier for the plants and for ourselves.

Unless you are prepared to accept a very high rate of propagation failure, at least some of the specialized equipment available for creating a protected environment is essential. The more seriously you become involved with plant propagation, the more spending on it will be justified.

Beginners often start by sowing seed in a tray under a sheet of glass, and by rooting cuttings enclosed in a plastic bag. But the constant need to wipe condensation from the glass or bag can become tedious. The next step is a simple heated propagator, with an electric element to provide bottom heat. This is the most efficient aid for germination and rooting. A further step leads to a sophisticated propagator, incorporating a misting unit which automatically controls the level of humidity. The difficulty with all these methods is to combine warmth and humidity with good ventilation, so that stems do not rot or leaves go mouldy. This is most likely to be achieved by using a misting unit in a warm greenhouse. The unit is not enclosed, inviting less danger from fungicidal diseases which quickly build up in still, stale, warm air.

Useful tools

A miscellany of equipment
1 Watering can
2 Hand mister for seedlings
3 Razor blades for leaf cuttings
4 Secateurs for taking hardwood cuttings
5 Pencil and labels
6 Presser to firm compost in seed trays
7 Trowel for planting out
8 Twine for tying
9 Garden line for marking out straight lines
10 Glass for covering seed trays
11 Sieve for removing lumps in compost/soil
12 Sharp knives to take and trim cuttings
13 Dibber to make holes for seedlings
14 Hormone rooting powder; fertilizer for indoor plants; fungicides and pesticides

Cold frame

Walls of brick, concrete or wood

12–18in
(30–45cm) high

8–12in
(20–30cm) high

Cold frames can be bought as kits or built fairly easily from 'Dutch lights', which have a single piece of glass. Two lights would make a frame 5ft/1.5m square. Erect the frame in shelter but not in shade.

Heated propagator

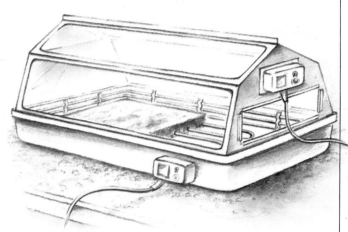

A propagator can be thought of as either a frame indoors or a miniature greenhouse. Like these, it is far more useful if heated. It costs much less to warm than a greenhouse and has the advantage over a frame of being indoors. Simple versions are fairly cheap to buy and to run, if fitted with a thermostat. A propagator used just for raising seeds does not need a tall cover but cuttings need more head room. It should have built-in ventilators to help prevent excessive condensation.

Mist unit

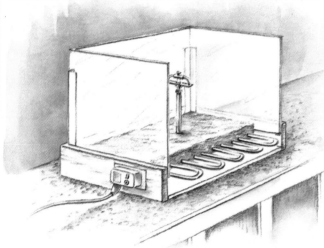

A mist unit will ensure far more success with seedlings and cuttings than any other device. It is not cheap, however, and needs to be used in a warm greenhouse. The unit is not enclosed, to allow plants adequate ventilation. It consists of a base with a thermostatically controlled heating element, maintaining a bottom heat around 70° F/21°C. The misters are fine nozzles raised above the base and connected to a water supply. They may be brought into operation in two ways. In some units a sensor increases the frequency of misting on sunny days. In others, an electronic 'leaf' activates a solenoid valve, which turns on the water until the required level of moisture in the air is reached.

Plastic bags and cloches

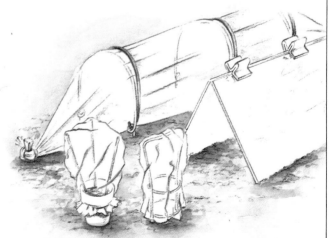

Enclosing a pot in a plastic bag creates a greenhouse effect. But note:
1 Put the bag over a wire hoop so that the plastic does not touch the leaves and make them rot;
2 Wipe condensation from the inside every day.
Plastic tunnel cloches protect hardy cuttings or growing plants. There are also cloches of rigid plastic: cheaper, lighter and more durable than glass, they deteriorate in the sun.

OUTDOOR PLANTS

Annuals

Some plants can expect to live, in a leisurely kind of way, for years or decades, but annuals have only one short season in which to grow, flower and seed. They compensate for the brevity of existence by filling it with astonishing exuberance throughout the summer months and into autumn. It is hard to exaggerate the fine performance of annuals, and at such comparatively low cost.

If you confine your choice to hardy annuals you will find they are also absurdly easy to grow: sown straight outdoors where they are to flower they will germinate in still coldish weather and survive spring frosts. Many will flower within weeks of sowing and with such prodigality as though there were no tomorrows — which for them is true.

This transience may have its drawbacks, but there is one great advantage for the gardener — the chance to recreate a new-look garden every year. This can be achieved merely by changing the plan of the beds, rearranging heights of plants, leaf patterns, colours, and so on. A garden planted solely with herbaceous perennials and shrubs can at times become too familiar to be fully appreciated, but you need never grow bored with a garden of annuals, especially if the garden is small.

Obviously, a garden need not be composed exclusively of annuals; they will blend well with

Seed sowing of annuals indoors can give young plants a head start. A propagator will provide the emerging seedlings with warmth and protection, while sowing in peat pots helps to lessen root disturbance when transplanting outdoors.

Sowing seeds outdoors

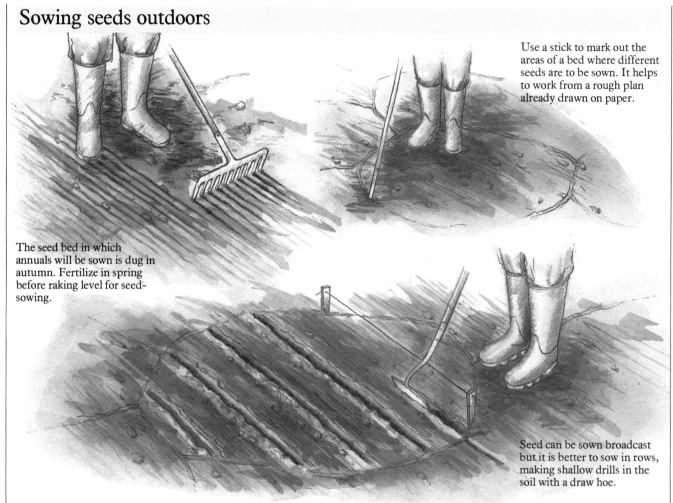

The seed bed in which annuals will be sown is dug in autumn. Fertilize in spring before raking level for seed-sowing.

Use a stick to mark out the areas of a bed where different seeds are to be sown. It helps to work from a rough plan already drawn on paper.

Seed can be sown broadcast but it is better to sow in rows, making shallow drills in the soil with a draw hoe.

herbaceous plants and provide brilliant patches of colour among more sober looking shrubs.

By adding to strictly hardy annuals a selection of half-hardy annuals and biennials — as well as perennials grown as half-hardy annuals — the range of colours and shapes and, above all, the season of flowering, are greatly extended.

Methods of cultivating annuals and other plants treated as annuals are explained in the following pages, together with a choice of popular plants.

TYPES OF PLANTS

True **annuals** are divided into hardy annuals (HA in the seed catalogues) and half-hardy annuals (HHA). Another category sometimes used is tender (T).

Hardy annuals are able to withstand some frost

SOWING EVENLY	SOWING UNEVENLY
Take a pinch of seed from your palm; scatter thinly.	Seed sown from the packet may fall out in a rush.

Sowing seeds in containers

Fill the seed tray generously with compost. Level to the rim with a piece of wood. Press down evenly with a flat piece of wood. After sowing (as shown on opposite page) sieve compost lightly and evenly over the seeds. Cover tray with a sheet of glass. On top of that place a sheet of newspaper (unless the seeds need light for germination). When germination begins, remove glass and paper.

in varying degree. The seeds are more hardy than the seedlings, which means that hardy annual seeds can be sown outdoors early in the year even though there may be a danger of slight frosts to come.

Half-hardy annuals can withstand some cold, but not sharp frosts. Therefore they can be sown outdoors only when all danger of significant frost has passed. Young seedlings will be damaged by even slight frost, so if the plants have been raised indoors they cannot be planted out until frosts are over and the soil has had time to warm up.

Tender annuals are so listed because neither the seeds nor seedlings can withstand frost at all.

Biennials grow over two seasons, producing leaves in the first and flowers and seed in the second, after which they die. But the cycle can be speeded up. In part this results from the development of new

hybrids, which, along with early sowing indoors, will produce flowers the same year, although later than if the plant had been grown the previous year. These plants are treated as though they were half-hardy annuals.

Perennials can live for many years, generally flowering each year, but some are grown as annuals. After these have flowered they are thrown away, for many perennials look better when young, growing straggly as they age and flowering less profusely. Perennials grown as annuals are treated as half-hardy.

GROWING HARDY ANNUALS OUTDOORS

To prepare the ground for sowing, dig well in advance the areas where the annuals are to be grown, preferably

Annuals

in autumn. This is also the time to dig in compost. Leave the ground rough for frost to crumble the surface during the winter months, while the soil underneath settles down. In early spring or spring, fertilizer can be scattered over the soil and lightly raked in, 10–14 days before sowing is planned. Be in no hurry in cold areas or if the season is late, for cold, wet soil is disastrous for germination.

Before sowing there are two final steps in preparing the bed. First, rake the soil to make it level and get rid of any lumps. Second, mark out with a stick the areas in which you want to grow the different plants. Avoid making each area too small or the whole bed when in flower will look like an over-bright jigsaw puzzle. It is best to have roughed out a design beforehand, working out on paper the vague ideas in your head. Points to take into account are the heights to which the plants will grow; times of flowering, so that there is a fairly even spread of colour throughout the season; and arrangement of colours to avoid clashes.

Within the overall design you can either sow the seeds broadcast or in rows. Broadcasting is a pretty lazy, hit-and-miss method. Sowing in rows gives better control over spacing the seeds, and after germination makes it easier to identify which are seeds and which weeds. It also simplifies thinning.

Allow 6in/15cm between the rows. Use string to get a straight line and take out a drill alongside it with a hoe or rake, or even a stick. The drills must be shallow: far too often annual seeds are sown too deeply. Fine seeds are best laid on the surface and then just pressed into the soil; larger seeds can be covered to a depth of two or three times their diameter. Sow thinly; outdoor sowing is often done far too thickly. To make the sowing of fine seed more even it may help to mix the seeds with sand, but the success of this ploy depends on how well the sand and seeds are mixed. Large seeds can be spaced at about the distance to which more thickly sown seed would first be thinned.

Some small seeds are sold pelleted (that is, coated to make them larger) but few annuals are treated in this way because of the cost of the process. Pelleted seeds are easy to space out at any distance you want.

After sowing, by whatever method, and when the seeds have been covered to the correct depth recommended on the packet, gently tamp down the surface of the soil. The flat side of a rake is useful for this. Then, before you forget, label the rows with the names of the seeds you have sown. This reminds you where you put them before they emerge, and later

Watering seeds

Seeds and young seedlings are best watered by immersion. Place the tray in shallow, slightly tepid water and leave there until drops of moisture creep up to the surface. Then remove the tray and let it drain.

helps you to recognise the various plants for what they are meant to be.

Germination times of plants vary by nature and because of the prevailing temperatures of both air and soil. Until the seedlings emerge and are growing well there is little to do except to keep the bed moist. Seedlings are very vulnerable at this stage because the roots are not long enough to reach down for moisture if the top soil dries out, either from sun or a drying wind. But water them gently, using a can with a very fine rose. Otherwise they will be washed right out of the ground, or flattened, and when that happens a seedling finds it hard to rise again.

As soon as the seedlings can be handled, thin them to around 6in/15cm apart, leaving the strongest seedling to grow. A later thinning may be necessary for larger plants, to the distances recommended on the seed packet, or in individual entries (pages 31–38).

GROWING HALF-HARDY ANNUALS

Sowings of half-hardy annuals outdoors are made later than sowings of hardy annuals; they cannot begin until all danger of frost has passed and the soil has begun to warm up. In cold areas this may mean that the plants are left with a short growing season and will be late in flowering.

There are several ways of cutting down the delay

Pricking-out and hardening off

A few days after pricking out, move trays of seedlings to a cold frame. Keep the frame lights shut at first, but over a couple of weeks or so, open them a little more each day.

To give growing seedlings more room they are pricked out when the first true leaves appear. Handle seedlings by their seed leaves — never by their stems.

Planting out

Seedlings are moved to the garden when the soil has begun to warm up. Use a dibber to make holes for the seedlings at the recommended distances.

Gently firm the soil around them and water them carefully from a can with a fine rose — they are still very fragile.

caused by waiting for good weather to arrive. The first is to germinate seeds indoors in pots or seed trays. Some seeds will need only gentle heat, to around 60°F/16°C, but others will need at least 70°F/21°C. If there are only a few seed containers, room might be found for them in the house — in the kitchen or a spare, but warm, bedroom. However, just as important as heat is adequate light, if the seedlings are to grow sturdy and not straggly. As a rule of thumb, start the seeds indoors four weeks or more before you can reasonably predict that the nights will be free from frosts, when the seedlings can be safely moved outdoors. Fine seeds take longer to germinate and should be sown a little earlier than large seeds.

Another method is to sow seeds in containers and put them in a garden frame. If you have a heated frame, sowings can be made as early as those indoors. For sowings in a cold frame you would need to wait a further two or more weeks.

SOWING HALF-HARDY SEEDS IN CONTAINERS, INDOORS OR IN HEATED FRAMES

Fill a seed tray or pot with seed compost, piling it up just above the sides. Press the compost down gently with your hands, not to firm it but to get rid of any air pockets, especially those which lurk at the corners of a

Annuals

tray. Level off the compost to the rim of the container with a piece of wood. Press the compost down evenly, not over firmly, with a flat piece of wood to just below the rim. The tray is now ready for sowing.

It, is unwise to sow seeds direct from the packet, because they tend to stick and then come out in a sudden rush. With very small seeds put some in the palm of one hand and sprinkle them on the compost from between thumb and finger of the other hand. Do not cover them with compost, just press them into the surface. Large seeds can be spaced out as you wish and covered lightly with sifted compost. Stick labels in the containers.

Water the compost either from above, using a can with a very fine rose, or, more effectively and with less disturbance to the seeds, from below. Stand the pot or tray in shallow tepid water and leave it for a time to let the water rise through the compost by capillary action. When the compost is damp throughout — little beads of moisture will appear on the surface — remove the container from the water.

Cover the tray or pot with a piece of glass, to keep in the moisture, and a sheet of newspaper to keep out too bright light. (But note that some seeds described on pages 31–38 need light to germinate, though not, of course, hot sunlight.) Keep a watch on the newly sown seeds and as soon as the seedlings appear, remove the glass and paper. Put the container in a good light, but not in the sun. Spray with water, making sure that the compost does not become soggy and compacted; the roots need air to grow strong and healthy.

PRICKING-OUT

The next operation, pricking-out, is most easily achieved when the first true leaves emerge, and do not delay long after that. The day before you intend to do the job, fill another container with fresh compost, water it, and keep it in the warmth overnight to avoid chilling the seedlings. The two implements useful for pricking-out are a forked stick to lift the seedlings and a dibber to make holes in the compost in which to replant them.

Loosen a few seedlings and lift one with the forked stick. Take hold of the seedling by a seed leaf, *never* by the stem. If you damage the stem the plant will never recover, but if seed leaves are damaged it matters little for they themselves will soon die. Make a hole with the dibber in the compost, lower the seedling into it, and firm the compost around it. Water in the seedlings and keep them in the warmth.

HARDENING OFF

When the seedlings have recovered from the shock of pricking-out, they can be hardened off. Begin by removing the containers to a cold frame. At first the lights of the frame must be kept shut altogether. At dusk on days when any frost is threatened, cover the frame with something to keep out the cold; a piece of thick old carpet will do. Soon, when the weather is suitable, wedge the lights open by day, a little at first. Later, during fine warm days, the lights can be taken off. After two weeks or so, when the seedlings have toughened up, the lights can stay off day and night in readiness for planting in the open. Keep these young plants moist, but not overwatered.

TRANSPLANTING

Transplant the seedlings to their prepared bed when the soil has warmed up a little. Space them at the distances recommended on the seed packet, or in the individual entries on pages 31–38. Water them in after transplanting.

SOWING HALF-HARDY ANNUALS IN UNHEATED FRAMES

Methods of raising seeds in unheated frames are the same as for heated frames, although sowings cannot be made as early. Germination will take longer and extra care will be needed to ensure that the frame is covered on frosty nights.

BIENNIALS AND PERENNIALS TREATED AS HALF-HARDY ANNUALS

These create problems in cold areas because of the consequences of late sowing. Growing biennials and perennials as half-hardy annuals means that they contract two growing seasons into one — and necessarily a long one. That is possible only with an early start indoors.

An obvious alternative is to treat the hardy biennials as biennials, among them *Dianthus barbatus* and *Campanula medium*. These are sown outdoors at various times from late spring to late summer in rows 12in/30cm apart, and pricked off into beds at 6–8in/15–20cm apart when large enough to handle. In autumn they are moved to their destined place in the garden, where they will grow to maturity and produce flowers in the spring and early summer of the following year.

Ageratum

SEED

Ageratum houstonianum 'Capri'. Half-hardy annual with fluffy flowers of bright blue, an unusual colour among annuals. This variety grows to only 6in/15cm, and is popular as a bedding plant.

Seed is sown indoors in gentle heat 2-2½ months before the likelihood of frost has gone and it will be safe to move the seedlings to the garden. Germination takes less than a week. Thin out the seedlings and harden off. Plant outdoors, 6-8in/15-20cm apart. Pinch out later to increase bushiness. Flowers bloom from early summer to summer (roughly 2-3 months from sowing) up until autumn frosts. Continued flowering depends on removing faded flowers; without that the plant is liable to stop.

Althaea

SEED

Althaea rosea, Hollyhock. Perennial usually grown as a biennial, and as an annual in some varieties. Grows to 6ft/1.8m, but there are dwarf forms.

Seed is sown sparingly in drills outdoors from late spring to early summer. Germination takes 2-3 weeks. Thin to 6in/15cm. Guard against slugs. Plant out in autumn, 18in/45cm apart, taking plenty of soil with the root. Seed can also be sown indoors from late winter to early spring to flower the same year, but flowers are later than, and inferior to, those grown as biennials.

Alyssum

SEED

Lobularia maritimum, Sweet Alyssum. Hardy annual used as a carpet or edging plant. Grows 3-4in/7.5-10cm tall, with honey-scented flowers of white, pink, purple, or lilac.

Seed is sown outdoors in early spring, broadcast thinly where plants are to grow. Sowing will be more even if seeds are mixed with a little sand. Rake in very lightly. At 60°F/18°C, germination takes about 14 days. Flowers appear 6-10 weeks later.

Antirrhinum

SEED

Antirrhinum majus, Snapdragon. Half-hardy perennial treated as a half-hardy annual. Depending on variety, it grows between 12in/30cm and 3ft/90cm.

Seed is sown from winter to early spring in heat of 65°F/18°C. To work out sowing time roughly, reckon 2 or 2½ months before it will be safe to put the plant outdoors. The seed is very small and barely needs covering; best results are likely with a peat based compost. Takes about 2 weeks to germinate. Prick out the seedlings and harden off before planting out when the danger of cold nights is over. Flowers bloom 4-5 months from sowing.

Aster

SEED

Callistephus chinensis, China Aster. Half-hardy annual. Heights vary from 6in/15cm to 30in/75cm and plants flower in innumerable colours.

Seed is sown from early spring to spring in a tray of peat seed compost. Barely cover the minute seed and keep at 65-70°F/18-21°C. Germination takes up to 2 weeks. Thin out and harden off. Plant out the taller varieties at 15in/37.5cm apart, the smaller rather closer. Flowers bloom from late July. For flowers over a longer period, make 2 or 3 successive sowings during early spring and spring. Alternatively, sow outdoors in late spring.

Begonia

SEED

Begonia semperflorens, fibrous rooted begonia. Half-hardy perennial, but treated as a half-hardy annual when used as a bedding plant outdoors. A bushy plant, 9-15in/22.5-37.5cm high with white, red or pink flowers.

Seed is sown from early to late winter. Prepare a seed tray with peat seed compost and sow on the surface. Press seeds gently on to the compost, but do not cover as they need light to germinate. Keep in a propagator at 70-75°F/21-24°C. Germination takes 2-4 weeks. Prick out when large enough and keep at a temperature of at least 60°F/16°C until planting out in June or July.

Calendula

SEED

Calendula officinalis, Pot Marigold. Hardy annual with vivid yellow and orange daisy-like flowers. Grows between 8in/20cm and 18in/45cm, according to variety.

Seed is sown outdoors in early spring or spring where plants are to bloom. Cover thinly. Germination takes up to 2 weeks. Thin out seedlings to 8in/20cm apart. Sow as early as weather permits so that plants flower before the summer grows too hot for them, when they are likely to stop flowering.

Campanula

SEED

Campanula medium, Canterbury Bell. Hardy biennial, unlike most campanulas, which are perennials. Seed must be sown every summer for the next year's flowers. The plant grows to 2-3ft/60-90cm, producing from late spring to early summer its bell-like flowers in blue, mauve, rose and white.

Seed is sown outdoors in late spring. Germination takes up to 3 weeks. Prick out to 4in/10cm apart when seedlings can be handled. These are not speedy growers, but should be ready for planting out in early autumn. They may need some protection through the winter.

Annuals

Centaurea

SEED

Centaurea cyanus, Cornflower. Hardy annual, easy to grow, reaching 12in/30cm or more, with rich blue flowers of purple, red, pink and white, etc. *Centaurea moschata,* Sweet Sultan, is another easily grown hardy annual. It reaches 18in/34cm, producing fluffy, scented flowers in many colours.

Seed is sown outdoors in early spring or spring, broadcast where plants are to flower. Gently rake into the soil. The seeds are a reasonable size and if sown carefully should need little thinning. Cornflowers do not object to some crowding. Germination takes 10–14 days and plants should flower about 2 months from sowing. Sweet Sultan should be sown broadcast outdoors in spring or late spring and the seedlings thinned to 9in/22.5cm apart.

Chrysanthemum

SEED

Chrysanthemum carinatum. Hardy annual, very easy to grow. The foliage is fern-like and the daisy-like flowers may be single or double, in gaudy colours. Grows to 24in/60cm.

Seed is sown outdoors, as soon as the weather is mild enough, in early spring or spring. Germination takes 1½–2 weeks, but double varieties may need longer. Thin to 12in/30cm apart. Flowers appear about 3 months from sowing.

Clarkia

SEED

Clarkia elegans. Hardy annual, with clusters of carnation-like double flowers in white, salmon, scarlet, purple, rose, etc. Grows to 24in/60cm.

Seed is sown outdoors where plants are to grow during early spring or spring, as soon as weather permits. Germination takes up to 2 weeks. Thin to 10in/25cm apart. Flowers bloom 3 months after sowing and continue, unless there is a very hot spell, until the first frosts.

Coreopsis

SEED

Coreopsis grandiflora. Hardy annual, easy to grow, with single daisy-like flowers in yellow, maroon and crimson on long stems with finely divided foliage. Tall varieties grow to 3ft/90cm; dwarf to 12in/30cm.

Seed is sown outdoors in early spring or spring; they can stand some degree of cold. Lightly rake in the seeds. Germination takes 2–3 weeks. Little thinning should be needed because some overcrowding seems to suit these plants. Flowers appear in profusion some 2–2½ months after sowing.

Delphinium

SEED

Delphinium ajacis, Larkspur. Hardy annual. Some varieties grow to 3ft/90cm; others to only 12in/30cm. Spikes of delphinium-like flowers in pink, lilac, blue or white are produced in summer.

Seed is sown outdoors in early autumn or early spring where plants are to flower. Germination takes 2–4 weeks. Thin to 10–12in/25–30cm apart. These plants are hard to transplant.

Dianthus

SEED

Dianthus caryophyllus, Clove Pink. Treat as a half-hardy annual. It grows 12–18in/30–45cm and produces very fragrant double flowers in white and shades of pink and red.

Seed is sown indoors from winter to early spring in pots or a tray filled with peat seed compost, at a depth of 0.25in /0.5cm.

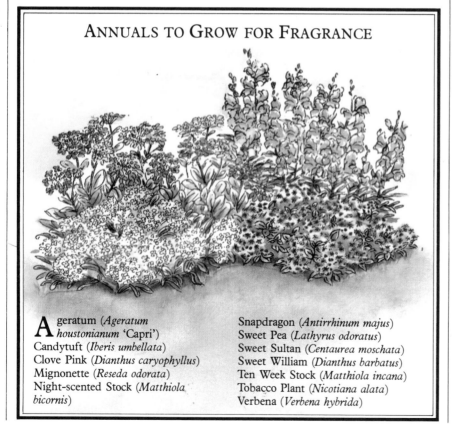

ANNUALS TO GROW FOR FRAGRANCE

Ageratum (*Ageratum houstonianum* 'Capri')
Candytuft (*Iberis umbellata*)
Clove Pink (*Dianthus caryophyllus*)
Mignonette (*Reseda odorata*)
Night-scented Stock (*Matthiola bicornis*)

Snapdragon (*Antirrhinum majus*)
Sweet Pea (*Lathyrus odoratus*)
Sweet Sultan (*Centaurea moschata*)
Sweet William (*Dianthus barbatus*)
Ten Week Stock (*Matthiola incana*)
Tobacco Plant (*Nicotiana alata*)
Verbena (*Verbena hybrida*)

Keep in a propagator at 60–65°F/16–18°C. Prick out when large enough to handle and plant outdoors in late spring, 8–10in/20–25cm apart. They need a sunny position and a non-acid soil.

Dianthus chinensis

Dianthus chinensis, the Indian Pink. Produces looser and rather less fragrant flowers than the Clove Pink. It grows 12–18in/30–45cm tall, with flowers on tall erect stems; good for cutting.

Seed of these pinks can be started indoors from winter to early spring and planted out from late spring. But as the plants grow quickly anyway, it is easier to wait and sow seed outdoors in spring where they are to flower. The seed is very small and barely needs covering. Thin to 6in/15cm apart. They will flower about 3 months after sowing and continue until the first autumn frosts.

Dianthus barbatus, Sweet William. Generally treated as a biennial, but some varieties can be grown as an annual. This popular cottage garden-type flower has scented flower heads of brilliant colours: crimson, scarlet and pink, marked with other colours and white. Grows 12–24in/30–60cm, or in dwarf varieties, 6–10in/15–25cm.

Seed is sown outdoors in a seed bed in late spring or early summer at a depth of 0.25in/0.5cm. Germination takes 2–3 weeks. Thin the seedlings to 6in/15cm

apart. In autumn transplant to flowering position, 8–10in/20–25cm apart. They will bloom the following year in early summer and summer.

Digitalis

SEED

Digitalis purpurea, Foxglove. Hardy biennial. Familiar as a wild flower, this plant grows to 4ft/1.2m and produces spikes of pink or, less often, white flowers. Perennial species have flowers in various shades of yellow.

Seed is sown in a seed bed outdoors in late spring to early summer. Do not cover the seeds as they need light for germination, which takes 2–3 weeks. In early autumn transfer seedlings to their flowering position, spacing them up to 12in/30cm apart. They will bloom the following year, from early to late summer.

Eschscholzia

SEED

Eschscholzia californica, Californian Poppy. Hardy perennial, but grown as a hardy annual. Reaches 12in/30cm, with lacy foliage and, usually, golden yellow flowers. Some forms have purple, cream or red flowers, however.

Seed is sown outdoors in early spring where plants are to flower. Germination takes 10–14 days. Thin to 8in/12cm. Flowers appear 6–8 weeks after sowing. These plants object to being transplanted.

Godetia

SEED

Godetia grandiflora. Hardy annual, easy to grow, with satiny, poppy-like flowers. Dwarf varieties reach 12in/30cm; others 24in/60cm.

Seed is sown outdoors in early spring or spring where plants are to flower. Germination takes 10–14 days. Thin to 8–12in/20–30cm apart, depending on the vigour of the variety. If overcrowded, they grow leggy, but given too much space they flower less well. Flowers bloom 3–4 months after sowing and continue for 10 weeks or more. The seedlings will not stand transplanting.

Gypsophila

SEED

Gypsophila elegans, Baby's Breath. Hardy annual which grows rapidly to 12–18in/30–45cm, producing a mass of tiny white or pink flowers on the numerous branches.

Seed is sown in early spring where plants are to grow. Scatter thinly and rake in lightly, firming the soil with a flat board. Germination takes up to 2 weeks. Thin to 10in/25cm or so apart. Flowers appear 6 weeks from sowing, but continue blooming for only 6 weeks. For successive blooms, make several sowings at intervals of 2–3 weeks.

Helianthus

SEED

Helianthus annuus, Sunflower. Hardy annual. The tallest varieties grow to 9ft/2.7m, and the smallest to 4ft/1.2m. Enormous bright yellow single flowers may be 12in/30cm or more across.

Seed is sown outdoors in early spring or spring where plants are to grow. Space seeds 3–4ft/90–120cm apart for the huge varieties; half this distance or less for the smaller and dwarf kinds.

Annuals

Iberis

SEED

Iberis umbellata, Candytuft. Hardy annual which grows, according to the variety, 6–12in/15–30cm high. Clusters of small flowers form a head which may be white or one of many bright colours.

Seed is sown outdoors at intervals of around 10 days from spring to midsummer to have plants in continuous bloom. Thin to 8in/20cm apart. Flowers appear 8–10 weeks from sowing.

Impatiens

SEED

Busy Lizzie. Most impatiens species now appear as hybrids under this familiar name. Although tender perennials, they are treated as half-hardy annuals when used as outdoor bedding plants. These very free-flowering plants bloom from early summer until struck down by frost and are found in endless colours. Most of the best are F_1 hybrids. They grow 6–15in/15–37.5cm.

Seed is sown indoors in early spring or spring in temperatures of 70–75°F/21–24°C. Cover seeds lightly and keep humid. Germination takes 2–3 weeks or more. Prick out and harden off in readiness for outdoor planting, which may not be possible until early summer. In autumn, before the first frosts, take cuttings to grow indoors.

Kochia

SEED

Kochia scoparia 'Trichophylla', Burning Bush. Half-hardy annual grown for its filmy pale green foliage which turns deep red in autumn. It grows to 24in/60cm.

Seed is sown indoors in early spring or spring, 6 weeks or so before the weather will be warm enough for planting out. Put seeds to soak in warm water 24 hours before sowing. Scatter on the surface of peat seed compost and place in a propagator at 70–75°F/21–24°C. Germination takes 7–15 days. Prick out

seedlings and gradually harden off for planting out, about 18in/45cm apart. Otherwise, when the weather is warm enough, sow outdoors in a sunny spot where plants are to grow. Thin to 18in/45cm.

Lathyrus

SEED

Lathyrus odoratus, Sweet Pea. Hardy annual growing to heights of 3–6ft/90–180cm and in a profusion of colours. Can also be treated as a half-hardy annual by being started off in gentle heat indoors in spring, or sown in autumn in a cold frame or greenhouse to overwinter and produce the earliest flowers.

Seed (the black variety) is sown after being soaked in water for a day or two. Alternatively, nick the tough seed coat with a file, but be careful not to damage the eye of the seed. Do not soak white seeds, which are liable to rot. Sow seed indoors between winter and early spring; judge when you expect the weather to be warm enough for planting out and sow 6–8 weeks beforehand. Keep the temperature between 55–65°F/13–18°C. Germination takes from 10 to 20 days. Weather permitting, harden off and plant out when the seedlings are some 4in/10cm high. Allow 6in/15cm between each plant. When 6in/15cm high, pinch out the growing tip to produce a bushier plant.
  Outdoors, sow in spring or late spring in deeply dug and well manured soil which has been allowed time to settle. Seeds should be 1in/2.5cm deep and 6in/15cm apart. Sow a few extra to fill any gaps. For overwintering, sow seeds in pots in early autumn or autumn and keep in a cold frame or greenhouse. Allow about 6 seeds to a 4in/10cm pot. Keep the compost just moist during the winter and plant out in spring.

Linaria

SEED

Linaria marrocana, Toadflax. Hardy annual, easy to grow, which is only

8in/20cm high. The flowers, like miniature snapdragons, are in many bright colours.

Seed is sown in early spring or spring where plants are to grow. Thin to 6in/15cm. Flowers appear about 8 weeks after sowing.

Linum

SEED

Linum grandiflorum rubrum, Scarlet Flax. Hardy annual, best planted in clumps for full effect. Grows to 12in/30cm.

Seed is sown outdoors in early spring or spring in a sunny part of the garden where plants are to grow. Germination takes 2 weeks. Thin to 8in/20cm apart. Flowers appear 3 months or more after sowing; successive sowings will continue the display into autumn.

Lobelia

SEED

Lobelia erinus. Half-hardy annual popular as an edging plant. It grows about 4in/10cm high and has a long flowering period with varieties in shades of blue, red, and white.

Seed is sown indoors in early spring or spring in a tray of peat seed compost, and started at a temperature of 65–75°F/18–24°C. Sowing should take place about 2 months before the weather in your area is likely to be warm enough to plant the seedlings out. Scatter the minute seed on the surface of the compost, which must be kept moist. Germination takes 1–2 weeks. The growing seedlings should not be kept too warm; gradually harden them off down to 60°F/16°C and plant out in the garden when still small, as soon as warm weather allows. Space 6in/15cm apart. Flowering begins 10–15 weeks after sowing and goes on and on.

Malcolmia

SEED

Malcolmia maritima, Virginian Stock.

Hardy annual and one of the easiest to grow. It reaches about 8in/20cm tall and produces a profusion of flowers in many colours.

Seed is sown outdoors in early spring or spring. Germination takes 7–10 days. Thin seedlings to 4in/10cm. Flowers may appear in 4 weeks.

Matthiola
SEED

Matthiola bicornis, Night-scented Stock. Hardy annual, very easy to grow, which reaches 10in/25cm. If mixed with seed of *Malcolmia maritima*, the results will give colour and scent by day and scent at night.

Seed is sown outdoors in early spring or spring. Germination takes 10–14 days. If sown in small patches in different parts of the garden, the evening air will be filled with its fragrance.

Matthiola incana, Ten Week Stock.

Treated as a half-hardy annual. Heights vary from 10in/25cm to 2ft/60cm, according to the variety. The plants are good for bedding and cut flowers. They bloom in many colours and have a clove-like scent.

Seed is sown from later winter to early spring in gentle heat of 60–65°F/16–18°C, and no higher. Time sowing for about 6–8 weeks before the weather is likely to be warm enough for transplanting outside. Germination takes 10–14 days. If double flowers only are required, choose strains which claim to be predominantly double and keep only seedlings with pale green leaves; dark-leaved seedlings will produce single flowers. The difference in leaf colour lasts rather longer if the temperature is fairly cool. So lower it to 50°F/10°C after the first leaves have appeared, to allow more time to distinguish the likely doubles. Plant out about 12in/30cm apart. Flowers appear 10 weeks or more from sowing. Seed may also be sown outdoors in late spring in areas where the plants are to flower.

Myosotis
SEED

Myosotis sylvatica, Forget-me-not. Hardy biennial with small flowers in many shades of blue, and some rose. Heights range between 5–12in/12–30cm.

Seed is sown outdoors from late spring to early summer to flower the following year. Germination takes 2–3 weeks. Transplant the seedlings to a cold frame in early autumn to overwinter. Move in early spring to flowering position, spacing plants about 6in/15cm apart.

Nemesia
SEED

Nemesia strumosa. Half-hardy annual with brightly coloured, small, orchid-like flowers. Grows to a height of 8–12in/20–30cm.

Seed is sown indoors from late winter to spring, about 9 weeks before it is likely to be warm enough to transplant outdoors. Start with a temperature of about 65–70°F/18–21°C. Prick out seedlings when large enough to handle and gradually harden off. When the weather is warm enough, transplant into the garden 6in/15cm apart. When plants are 6in/15cm tall, pinch out tips to increase bushiness.

Nicotiana
SEED

Nicotiana alata, Tobacco Plant. Perennial grown as a half-hardy annual. It reaches about 30in/75cm and produces tube-shaped flowers with petals, like a five-pointed star, carried on long stems. Colours include white, mauve and maroon. Blooms are strongly scented in the evening.

Seed is sown indoors, 8–10 weeks before it will be safe for seedlings to be transplanted outside. Do not cover the minute seeds, just press them into the compost. When planting out, space seedlings 24in/60cm or more apart. Flowers bloom from summer onwards.

DAMPING-OFF

Apparently healthy seedlings may suddenly rot at ground level, collapse and die. This is known as 'damping-off' and the cause is a soil-borne fungus disease. Once a seedling has collapsed nothing can be done to rescue it, but some precautions can be taken to prevent attacks.

Always use sterilized compost for seeds sown indoors. Keep seed trays and pots absolutely clean.

Buy seed which has been treated with fungicide. Or do this yourself by adding a little powdered sulphur to the packet of seeds and shaking the contents together. This will protect the seeds for a short time only, until roots reach down into the diseased soil. So water the compost after sowing with a dilute solution of one of the organic systemic fungicides on sale and use the same solution to spray emerging seedlings.

Wet compost and a humid atmosphere encourage damping-off. Avoid over-watering, and if the seed tray or pot is covered with a sheet of glass, wipe the underside every day to prevent condensation from dropping on to the seedlings.

Seedlings huddled closely together are far more liable to attack. Sow thinly.

Annuals

Nigella

SEED

Nigella damascena, Love-in-a-Mist. Hardy annual. Most varieties grow to 18in/45cm; they produce lacy foliage and blooms reminiscent of cornflowers, in blues, pink and white.

Seed is sown outdoors in spring or late spring. Germination takes about 2 weeks. Thin out to 6in/15cm. Flowers appear 10–12 weeks from sowing. They are not easy to transplant.

Papaver

SEED

Papaver rhoeas, Shirley Poppy. Annual. Sophisticated version of the cornfield poppy, available in a variety of colours. Grows to 2ft/60cm.

Seed is sown outdoors in early spring or spring, where plants are to flower; they hate to have their roots disturbed. Scatter seeds sparingly, an operation which is easier if they are mixed with a little sand. Germination takes 10–14 days. Thin seedlings if necessary, enough to avoid overcrowding. Flowers bloom 8–10 weeks from sowing.

Penstemon

SEED

Penstemon × gloxinioides. Hardy perennial, but treat as a half-hardy annual and it will flower the first year from seed. Most penstemons grow to under 2ft/60cm; their flowers are like those of foxgloves, but smaller, in colours from white, through pink, to crimson.

Seed is sown in winter or late winter in a temperature of 55–65°F/13–18°C. Sprinkle seeds on the surface of the compost, since they need light for germination, which takes 20 days or longer. Harden off seedlings in a cold frame and plant out when there is no more danger of frost. Space 12in/30cm apart.

Petunia

SEED

Petunia grandiflora

Petunia grandiflora; Petunia multiflora. Half-hardy annuals. *Grandiflora* has large flowers; *Multiflora* has a large number of small flowers. Many of the older varieties have been eclipsed by the F_1 hybrids, which have single or double flowers from all colours of the spectrum, and wavy, fringed or textured petals. Most grow 8–12in/20–30cm.

Seed is sown indoors from winter to early spring. Temperatures of 75–80°F/ 24–27°C are required in the propagator. A reasonable guide to sowing time allows 2 months or more before it will be safe to move the young plants outdoors. These seeds are minute, so sow with care, especially the expensive seeds of F_1 hybrids. Germination takes up to 2 weeks. When pricking out seedlings do not reject the smaller ones — they are likely to produce double flowers or have the most vivid colours. Plant outdoors at 8–12in/20–30cm apart and pinch out growing tips when the plants are about 6in/15cm tall, to encourage bushiness.

Phlox

SEED

Phlox drummondii. Half-hardy annual, a popular bedding plant, producing a profusion of many-coloured flowers over a long period. Dwarf varieties grow to about 6in/15cm; taller varieties to 12in/30cm.

Seed is sown indoors at around 65°F/18°C in early spring and plants should be ready for planting out 6–8 weeks later. Germination takes 10–14 days. Transplant with as much compost round the roots as possible, for the plants dislike disturbance. Space 10–12in/25–30cm apart. Flowers appear 8–10 weeks from sowing and continue blooming into autumn. Seed can also be sown outdoors when the weather is warm enough, probably in late spring.

Reseda

SEED

Reseda odorata, Mignonette. Hardy biennial, but treat as a hardy annual. This plant is not showy and tends to sprawl, growing 12in/30cm tall, but is grown for the pleasing fragrance of its greenish-yellow flowers.

Seed is sown outdoors in early spring or spring. The seed is minute, should be sown only thinly, and lightly covered. Germination takes 10–14 days. Thin to 8in/25cm apart; it does not transplant at all well. Flowers appear 8–10 weeks from sowing. A succession of sowings every 2 weeks until summer will provide flowers into the autumn.

SOWING FOR SIZE

The height to which a plant will grow depends in the first place on the variety chosen; there are even miniature sunflowers. When buying seeds, check carefully that you are choosing the type you want. The height of a plant is also affected by its environment, such as a fertile or impoverished soil and fine or bad weather. Height forecasts can therefore be of only rough guidance.

TALL ANNUALS FOR BORDER BACKGROUND

Althaea rosea (Hollyhock)
Delphinium ajacis (Larkspur)
Digitalis purpurea (Foxglove)
Helianthus annuus (Sunflower)
Lathyrus odoratus (Sweet Pea)
Rudbeckia hirta (Gloriosa Daisy)
Tagetes erecta (African Marigold)

MID-BORDER ANNUALS

Antirrhinum majus (Snapdragon)
Calendula officinalis (Pot Marigold)
Centaurea moschata (Sweet Sultan)
Clarkia elegans
Dianthus barbatus (Sweet William)
Godetia grandiflora
Kochia scoparia (Burning Bush)
Nicotiana alata (Tobacco Plant)
Nigella damascena (Love-in-a-Mist)
Papaver rhoeas (Shirley Poppy)
Scabiosa atropurperia (Sweet Scabious)
Zinnia hybrids

SMALL ANNUALS FOR BORDER FOREGROUND

Ageratum houstonianum
Begonia semperflorens
Dianthus chinensis (Indian Pink)
Iberis umbellata (Candytuft)
Impatiens hybrids
Lobelia erinus

Lobularia maritimum (Sweet Alyssum)
Myosotis sylvatica (Forget-me-not)
Nemesia strumosa
Petunia hybrids
Salvia splendens (Scarlet Sage)
Verbena hybrids
Violas
Zinnia hybrids

Rudbeckia

SEED

Rudbeckia hirta, Gloriosa Daisy. Treat as a half-hardy annual. The plant grows to 3ft/90cm, and its golden and red daisy flowers may reach 6in/15cm in diameter. Other species are smaller.

Seed is sown outdoors in spring or late spring. Germination takes 2–3 weeks. Space 2ft/60cm apart.

Salvia

SEED

Salvia splendens, Scarlet Sage. Hardy perennial treated as a half-hardy annual. Grows 6–18in/15–45cm tall, according to variety.

Seed is sown early, as salvias are slow growing. Begin in late winter or early spring in a propagator at 70–80° F/ 21–27° C. Soak seeds for a day to help germination, which takes 10–14 days. Seedlings resent being transplanted, so sow in peat pots to avoid disturbing the roots. Move outdoors only when the weather has turned warm. Plant dwarf varieties 12in/30cm or so apart; taller varieties 30in/75cm. They must have full sun. Flowers bloom 10–12 weeks from sowing.

Scabiosa

SEED

Scabiosa atropurperia, Sweet Scabious. Hardy annual. Long, slender stems, 2ft/60cm high, bear a pincushion of blue flowers.

Seed is sown outdoors in spring or late spring in a sunny part of the garden. Germination takes 3 weeks. Thin to 8in/20cm apart. To encourage bushier growth, pinch out tips of branches. Flowers bloom about 12 weeks from sowing, and with regular dead-heading will continue until the first frost.

Tagetes

SEED

Tagetes erecta, African Marigold. Half-hardy annual. Dwarf varieties grow to 12in/30cm; taller varieties to 3ft/90cm. Flowers may be gold, yellow and orange. Many are F_1 hybrids.

Seed is sown in gentle heat indoors from late winter to early spring, about 8 weeks before the plants can safely be moved outdoors. Germination takes little more than a week and transplanting is no great setback. Space tall varieties 18in/45cm or more apart; the dwarfs only 12in/30cm. Seed may also be sown directly outdoors in a warm position in early spring. Early sown seeds will flower within 2 months, but flowers from spring sowings may be 2 weeks later. Pinch out to produce bushier plants and more profuse flowering.

Tagetes patula, French Marigold. Half-hardy annual. Cultivate as for the African Marigold, but with closer planting.

Tagetes tenuifolia pumila (syn. *Tagetes signata pumila*), Tagetes. Half-hardy annual, growing up to 8in/20cm, with small domes of golden flowers.

Seed is sown in gentle heat from late winter to early spring, or outdoors in spring. A long flowering period extends from end of early summer.

Tropaeolum

SEED

Tropaeolum majus, Nasturtium. Hardy annual. Varieties may be dwarf and bushy (6–9in/15–22.5cm), semi-trailing (12in/30cm), or climbing (to 6ft/1.8m).

Seed, which is quite large, is sown outdoors where plants are to grow in spring or late spring. Sow dwarf varieties 10in/25cm apart; others about 2ft/60cm apart, and at a depth of 3 times the diameter of the seeds. Germination takes up to 2 weeks.

Verbena

SEED

Verbena hybrida. Half-hardy when grown as a plant for bedding. Varieties of these spreading and trailing plants are mostly no more than 8in/20cm high. Clusters of fragrant small flowers are produced in a variety of bright colours.

Seed is sown from winter to early spring in gentle heat. Germination is often slow, taking up to 3 weeks; keep the compost just moist, not wet. Prick out into trays when large enough to handle. Transplant outdoors only when warm weather has arrived, choosing a spot in full sun. Space 12in/30cm or more apart. Flowering begins 12–15 weeks from sowing.

Viola

SEED

Viola × wittrockiana, Pansy, or Heartsease. Half-hardy annual. This viola hybrid grows to 8in/20cm, but sprawls.

Seed is sown from late winter to early spring in gentle heat of up to 60°F/16°C. Germination takes 2–3 weeks. Never use old seed, that is, over 9 months old, as this germinates badly; even fresh seed will not germinate above 70°F/21°C. Plant outdoors when danger of serious frost has passed, spacing seedlings 6in/15cm apart. Flowering starts from late spring to early summer. Plants for sale already in flower in early spring have been sown the previous autumn. For early flowers, sow in late summer and overwinter in a cold frame or cool greenhouse at 45–50°F/8–10°C. Plant out in early spring, 6in/15cm apart.

Viola cornuta, Viola. Perennial but treated as a half-hardy perennial. This typical cottage garden variety is more modest than the blousier ostentatious *wittrockiana*, with smaller but very profuse blooms.

Seed is sown in gentle heat in late winter to early spring. Plant out, 4in/10cm apart, when frosts have gone.

Xeranthemum

SEED

Xeranthemum annuum. Hardy annual, easily grown, with daisy flowers that can be dried for the winter. Reaches 2–3ft/60–90cm.

Seed is sown outdoors in early spring or spring where plants are to grow. Germination takes 2–3 weeks. Thin to 8in/20cm apart. Flowers appear up to 3 months from sowing.

Zinnia

SEED

Zinnia hybrids. Half-hardy annuals. These plants may be tall (30in/75cm) or dwarf, with single, semi-double or double daisy-like flowers in most colours except blue.

Seed is sown indoors in early spring or spring. Use peat pots since these plants cannot stand transplanting. Allow one seed to a pot. Start off in a propagator at 75°F/24°C and then lower to not less than 60°F/16°C a few days after germination. This may take anything from 10 days to 3 weeks. Avoid a too humid atmosphere while seedlings are growing. Plant out when all danger of frost has gone. Space tall varieties 12in/30cm or more apart, and the dwarfs 8–10in/20–25cm. Flowering begins 8–10 weeks after sowing. Seed may also be sown outdoors in late spring.

Annual Climbers

Some perennial climbers can be treated as half-hardy annuals to give rapid results in covering walls or pergolas. They will flower the same year, but need to be started off in heat.

Calonyction

SEED

Calonyction aculeatum, Moonflower, tender plant growing 15ft/4.5m or more, with large scented pink and white flowers (opening at dusk) from summer until the first frosts.

Seed is sown in spring. Soak the hard seeds overnight or file the seed coat. Sow in peat seed compost, only lightly covering the seed. Keep in a propagator at 70°F/21°C, pricking out the seedlings and later hardening them off until planting outdoors after all danger of frost is gone.

Cardiospermum

SEED

Cardiospermum halicababum, Balloon Vine, a perennial vine grown as a half-hardy annual. It rapidly reaches 10ft/3m, with tendrils which need wires to cling to. Small white flowers appear in summer among a mass of feathery foliage. The seed pods which follow are like small papery balloons, about 1in/2.5cm in diameter.

Seed is sown outdoors in late spring, when the weather has turned warm. Choose a spot where the plants are in full sun. Thin out the seedlings to 18in/45cm apart. Keep the plants moist at all times. They will be killed by the first frosts.

Cobaea

SEED

Cobaea scandens, Cup and Saucer

Plant, a half-hardy but rampant climber (with tendrils) which can grow to 20ft/6m. From late summer to autumn appear the violet, bell-shaped flowers (the cup) surrounded by greenish bracts (the saucer).

Seed is sown indoors in spring at 65°F/18°C. Allow one large flat seed to a pot, pushed sideways into the peat seed compost. Germination takes 2 weeks. When warm enough, plant out some 3ft/90cm apart.

Ipomoea

SEED

Ipomoea rubro-caerulea or *tricolor*, Morning Glory. Half-hardy plant with soft-blue, trumpet-shaped flowers in summer, opening at sunrise and fading by mid-afternoon except when cloudy.

Seed is sown in peat seed compost and kept at 70°F/21°C. Soak the seed the day before sowing or file the hard coat. Prick out and harden off before planting outdoors when the weather is warm enough. Space 18in/45 cm apart.

Quamoclit

SEED

Quamoclit coccinea, Star Glory, a self climber growing more than 10ft/3m, with clusters of fragrant scarlet flowers from early summer to early autumn.

Seed is sown outdoors when there is no danger of frosts. File the hard coat of the seed and then soak in hot water and leave overnight. For earlier flowers, sow indoors 6 weeks in advance and plant out after hardening off. Space 3ft/90cm or so apart.

Thunbergia

SEED

Thunbergia alata, Black Eyed Susan, a half-hardy twining climber growing to about 4ft/1.2m. From summer onwards, it has white, yellow or orange flowers with a black 'eye'.

Seed is sown outdoors where plants are to bloom when soil and weather are warm, or indoors 6 weeks earlier. Sow 2 or 3 seeds in a 3in/7.5cm pot of peat seed compost and keep at 65°F/18°C. Germination takes 2–3 weeks. After hardening off, plant out in sun.

Tropaeolum

SEED

Tropaeolum peregrinum, Canary Creeper, is a genuine annual and not a perennial treated as a half-hardy annual. Grows 6ft/1.8m or more, and needs strings to cling to. Pale green leaves on slender stems and many lemon-yellow flowers bloom from summer.

Seed can be sown outdoors, but wait until the soil has warmed up. Cover the seeds to a depth of 3 times their size. Final spacing should be 2ft/60cm.

Herbaceous Perennials

Herbaceous perennials have a longer, often much longer, life than annuals. They give a more permanent feeling to a garden, even though their leaves and stems die down during the months of winter. The great days of the grand herbaceous border, which flourished at the start of this century, are now almost gone. Although magnificent to look at, this feature has two drawbacks: it takes an immense amount of time to look after, and, like a garden of annuals, is for summer-time only, with a slight overlap into autumn. A garden border nowadays is most likely to be a mix of herbaceous perennials and shrubs, plus bulbs for early and late flowers and annuals for added summer brilliance.

Among the herbaceous perennials are many which were cottage garden flowers when cottage gardens existed in reality, and not just in pictures. Many of these plants have been 'bred up' in the world, to become costly F_1 hybrids. Some are improved by this treatment, but others become a parody of the humble originals. The amateur propagator should beware of being seduced by the word 'new' in seed catalogues.

PROPAGATION METHODS

Herbaceous perennials offer far more opportunities for propagation than annuals. Not only can they be raised from seed (by sexual propagation) but by vegetative (or asexual) propagation as well — using bits of stems and roots to create identical reproductions of themselves.

Because the plant stems of perennials die down each year, their stem cuttings are mainly the softwood

Contrasting areas of colour in the garden can be provided by a careful selection of herbaceous perennials. Such reliable old favourites as *Dianthus* (pinks) and *Kniphofia* (Red Hot Poker) can be kept at their best by renewing older plants every few years.

variety. But the roots live on from year to year and because of the great variety of rooting systems — including fibrous roots, rhizomes and tubers — many forms of propagation are possible. The simplest, as almost always, is by division.

HOW TO DIVIDE

Plants can be divided at any time in suitable weather when the plant is dormant, from late autumn to late winter, but there is much to be said for waiting until early spring when the plant is starting into growth. The soil will then be warming up gradually, encouraging more rapid new root growth. Dig up the plant carefully with a fork. If it is small enough, pull it in two by hand. Large clumps can be separated by using two garden forks thrust back to back into the thickest part of the clump. By levering the forks apart, the plant should break into two. Remove and throw away the centre growth of the plant, which is the oldest. Keep the outer, younger growth.

Further division may also be possible, if each part has a number of buds or shoots with plenty of fibrous roots to sustain them. How savagely plants can be divided, and how soon they will recover, varies from plant to plant. As a general rule, reckon that the divided clumps should be at least 4in/10cm across. In this, as so often, experience is the teacher, helped by luck. Replant immediately. Water, and in a dry spell, keep watered.

Some fibrous-rooted plants can be divided into very small sections; the stems, each with roots attached, almost fall to pieces in your hands. This form of division is known as **'Irishman's cuttings'** (page 17). The 'cuttings' are replanted straight away. Plant fairly closely so that the new growth forms a clump in a reasonable time. Michaelmas daisies are suitable subjects. **Divided tubers** can be used to make new plants; this is a method for dahlias (page 46).

Some herbaceous perennials cannot be divided because they do not form clumps. These include lupins, Oriental poppies, carnations and pinks. They must be propagated in other ways. It is also unwise to divide some plants because they hate root disturbance so much. Ignore this and they may take reprisals by not flowering for several years.

ROOT CUTTINGS

Roots of some plants can be propagated from cuttings as well as by division; one such is *Papaver orientale*. Take root cuttings when the plants are dormant. Small plants may be dug up, but with larger plants, try removing some soil from them first to see whether they will provide suitable cuttings; the young roots you need are likely to be around the edge. Cut away only a few roots, replant the parent or replace the soil around it, and then prepare the cuttings. Trim the roots into pieces 2in/5cm long; the top cut is made straight across the root while the bottom cut is slanted.

Dividing roots

Separate stubborn clumps by prising apart with two garden forks. This does less damage than chopping with a spade.

Dividing tubers

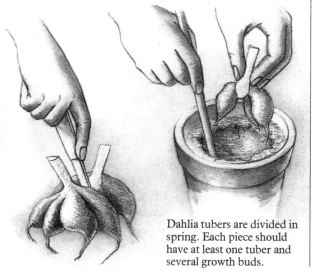

Dahlia tubers are divided in spring. Each piece should have at least one tuber and several growth buds.

Push the cuttings vertically into the compost, slanting end down as on the original root, until just level with the surface. Cover with a shallow layer of compost or grit. (Some gardeners plant thin cuttings horizontally, but fewer root that way.) Cuttings from hardy plants can go into a cold frame. Keep the lights shut all winter and leave them there until autumn, when they can be planted out in the garden or moved to a nursery bed. Half-hardy root cuttings raised in heat must be well hardened off before being moved.

STEM CUTTINGS

Ordinary **softwood cuttings** are the trickiest stem cuttings to root. The best chance of success comes from **basal cuttings**, which are very young shoots growing from the crown of the parent plant. Dahlias are an example. The tubers are brought into the warmth in winter or late winter to force them into growth. When the shoots are some 2in/5cm long they are removed, with a small 'heel' of the tuber attached. Plant in warmth in a mixture of moist peat and sand and they will root in a few weeks. Move into individual pots and harden off until they can be planted outdoors.

The problem with softwood cuttings taken later is that they soon wilt. Plant them quickly and keep in humid warmth but maintain good ventilation. A heated propagator will generally meet these conditions; a mist unit would be even better.

In late spring or early summer, take cuttings 4in/10cm long. Do not remove the tip of a softwood cutting, but trim the bottom end to just below a node. Remove the lower leaves, so that a third or half of the cutting is bare. (Too many leaves left on the cutting above the compost cause rapid loss of moisture; leaves buried in the compost induce rot.) Dip the end of the cuttings in hormone rooting powder and plant them almost to the depth of the bare stem in pots or trays of the recommended compost. Temperatures at which hardy plants are kept to root are generally lower than for houseplant cuttings.

Semi-ripe cuttings are taken between summer and early autumn. The method is similar as for softwood cuttings, except that the tips of the cuttings are removed. Few herbaceous perennials are propagated from semi-ripe cuttings, however.

SEED

Seeds of herbaceous perennials are sown indoors if they are not thoroughly hardy, but otherwise outdoors. The methods of sowing are as for annuals or half-hardy perennials treated as annuals, and are explained on pages 26–30. Treatment starts to differ with the after care of the young plants, which will usually begin to flower in their second year. Those sown indoors are moved, after hardening-off, to cold frames or a 'nursery bed'. They are then moved to their permanent home around early autumn or in early spring the following year.

Root cuttings

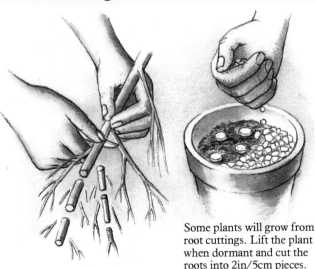

Some plants will grow from root cuttings. Lift the plant when dormant and cut the roots into 2in/5cm pieces.

Basal cuttings

Basal cuttings are the very young shoots growing from a plant crown. They are one way of propagating dahlias.

Achillea

DIVISION; SEED

Grows to 4ft/1.2m. *Achillea filipendula*, Fernleaf Yarrow, produces large, flat, circular heads of densely- packed, bright yellow flowers in summer. The mid-green, fern-like leaves are deeply cut.

Division takes place in spring. But allow plants to become well established before dividing every 3 to 4 years. Replant in a sunny position, 18in/45cm apart.
Seed is sown in late spring outdoors in a seed bed, lightly covering with compost. Germination takes 4–12 weeks. Plant out to flowering positions in autumn, about 18in/45cm apart to allow for spread of plant. Earlier sowings can be made under glass from late winter onwards at 60–65°F/16–18°C.

Aconitum

DIVISION; SEED

Grows to 4ft/1.2m. *Aconitum carmichaelii*, Monkshead, produces spikes of hooded, violet-blue flowers in late summer, with dark green, leathery, deeply incised leaves. *Aconitum napellus* has similar flowers, of deep blue, in summer, and deeply cut leaves.

Division takes place in spring. Lift and divide clumps of tuberous roots into pieces. Take care when handling the plants as they are poisonous. Plant 12in/30cm apart. Will tolerate a part shaded

Seed is sown outdoors in spring, just covering the seed with compost. Plant out to flowering positions in autumn, 12in/30cm apart. Seed can also be sown in trays in gentle heat, 55–60°F/ 13–16°C. Do not cover seed, as light aids germination. Plant out in autumn.

Agapanthus

DIVISION; SEED

Agapanthus africanus

Agapanthus africanus, African Lily, has a rounded head of deep blue to violet lily-like flowers carried on a 3ft/90cm stem in late summer. Strap-shaped, leathery leaves die down in winter. May need winter protection in all but mild areas. 'Headbourne Hybrids', developed mainly from *Agapanthus campanulatus*, produce in summer large heads of pale to deep blue flowers. Its strap-shaped leaves grow to 3ft/90cm.
Division takes place in late spring or early summer. Lift and divide the rhizome, ensuring each piece has several growing eyes and plenty of roots. Plant 18in/45cm apart in a warm, sunny position.

Seed is sown in spring. Sow seed singly in 3in/7.5cm pots of loam seed compost. Place in a propagator at 60–65°F/ 16–18°C. Germination is erratic, taking 6 weeks to 3 months. Grow for 2 years in a greenhouse or frame and then plant out 18in/45cm apart in late spring. Takes up to 5 years to reach flowering size.

Alchemilla

DIVISION; SEED

Grows to 18in/45cm. *Alchemilla mollis*, Lady's Mantle, has large, rounded, downy, grey-green leaves and feathery heads of yellowish-green flowers in summer. It self-sows easily and can be troublesome. Pick off flower heads before seed develops.

Division takes place in spring. Lift plants and divide the roots, replanting 18in/45cm apart in sunny or shaded position.
Seed is sown in spring. Fill a pan with loam seed compost and scatter seed evenly over the surface. Just cover with compost. Place in a propagator at 60°F/16°C. Germination takes 6 weeks. Harden off and grow in a cold frame until autumn or following spring. Plant out 18in/45cm apart.

Aquilegia

SEED; DIVISION

Grows to 3ft/90cm. *Aquilegia alpina* produces 12in/30cm stems bearing nodding flowers of deep blue in late spring and early summer. *Aquilegia hybrida*, Columbine, has white, yellow, blue, purple, pink or bright red flowers.

Seed is sown in early spring in a tray of loam seed compost, just covering the seed. Place in a propagator at 65–70°F/ 18–21°C. Germination takes 4–12 weeks. Thin out seedlings when they can be handled easily and harden off. Plant out in summer, 12in/30cm apart. Outdoor sowings can be made from late spring to early summer in a prepared bed. Sow seed 0.25in/0.5cm deep. Thin out and eventually transfer to flowering positions in early autumn, 12in/30cm apart. Seed of *Aquilegia alpina* is better sown in pots in autumn and left outside during winter to subject the seed to frost. Germination takes place the following spring.
Division of species other than *Aquilegia alpina* takes place in spring. Lift and divide, replanting 12in/30cm apart. Does not object to partial shade. *Aquilegia alpina* dislikes disturbance.

Aster

SEED; DIVISION

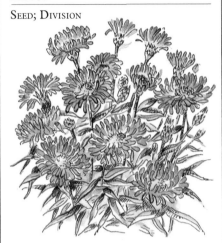

Grows to 4ft/1.2m. *Aster novi-belgii*, Michaelmas Daisy, has daisy-like white, red, pink or purple flowers, double and semi-double, in late summer and early autumn. Tall growing forms require staking but dwarf varieties grow only 6–12in/15–30cm tall. The leaves are lance shaped.

Seed is sown in trays of loam seed compost in spring. Cover seed to its depth with compost. Place in a propagator at 60–65°F/16–18°C. Germination takes 2 weeks. Thin out seedlings, harden off and transplant outdoors to flowering positions in late spring at 12in/30cm intervals. Outdoor sowings in beds can be made from late spring to summer. Thin out to 12in/30cm.
Division takes place in spring every 3 to 4 years. Divide large clumps into several pieces, each with young, sturdy growth: Replant 12in/30cm apart.

Aubretia

SOFTWOOD CUTTINGS; DIVISION; SEED

Grows to 6in/15cm. *Aubretia deltoidea*, Purple Rock Cress, has mats of star-like spring flowers of deep blue, purple or pink.

Softwood cuttings are taken in late spring and summer. Remove tip cuttings, about 2in/5cm long. Strip off the bottom leaves and plant in a mixture of two thirds loam potting compost and one third sand.

Place in a propagator at 65°F/18°C. When new growth appears, harden off and plant out in early autumn.
Division takes place in early spring. Lift large mats and divide, replanting 18in/45cm apart.

Seed is sown in spring. Fill a tray with loam seed compost and scatter seed evenly over the surface. Lightly cover with compost. Place in a propagator at 65–70°F/18–21°C. Germination takes 3 weeks. Thin out seedlings when they can be handled easily. Harden off and transplant to flowering positions, 18in/45cm apart, in early autumn. Seed can be sown outdoors in a seed bed from late spring to early summer. Cover to a depth of 0.25in/0.5cm. Thin out seedlings to 6in/15cm when they can be handled easily. Transplant to flowering positions, 18in/45cm apart, in early autumn.

Campanula

SEED; DIVISION; SOFTWOOD CUTTINGS

Campanula carpatica

Grows to 2ft/60cm. *Campanula carpatica*, Bellflower, has white, blue or purple bell-shaped flowers in summer. The small, closely-packed, light green leaves are rounded. It grows to 9in/22.5cm. *Campanula persicifolia*, Peach-leaved Bellflower, has white or blue, bell-shaped flowers on tall 2ft/60cm stems in summer, and dark green, narrow, peach-like leaves.

Seed is sown in spring. Fill a tray with loam seed compost and press seed gently into it. Place in a propagator at

60–65°F/16–18°C. Germination takes 4 weeks. Prick out seedlings when they can be handled easily. Harden off and plant out in flowering positions in early autumn, 12in/30cm apart. Outdoor sowings can be made in late spring. Cover seeds to a depth of 0.25in/0.5cm. Thin out when they can be handled easily and transplant to flowering positions in early autumn, 12in/30cm apart.
Division takes place in spring, every 3 to 4 years. Lift and divide into several pieces, replanting 12in/30cm apart.
Softwood cuttings are taken in spring. Remove 3in/7.5cm stem tip cuttings and strip off the lower leaves. Plant in a mixture of two thirds loam potting compost and one third sand. Place in a propagator at 60°F/16°C, or under glass in a greenhouse or cold frame. When new growth appears, harden off and plant out.

Canna

DIVISION; SEED

Grows to 4ft/1.2m. *Canna indica*, Indian Shot, has large, leathery, oval to lance-shaped leaves ranging in colour from green to purple. Yellow, orange, red or pink flowers bloom in summer. Dwarf forms growing to 18in/45cm.

Division takes place in spring, the tubers having been lifted after flowering and stored in sand in a frost-free place during winter. In spring, divide the tubers, ensuring that each piece has one or more growing points. Plant in boxes of loam potting compost and place in a greenhouse or frame at a minimum 60°F/16°C. When leaves appear, harden off and plant out to flowering positions in late spring, 2ft/60cm apart, when all danger of frost is over.

Seed is sown in spring. Soak in tepid water for 24–48 hours. The seed surface can also be chipped to aid germination, but take care not to damage the embryo inside. Sow the large seed individually, in 3in/7.5cm pots of loam seed compost, covering to the depth of the seed. Place in a propagator at 70–75°F/21–24°C. Germination takes 8 weeks. Harden off and plant out in early summer, 2ft/60cm apart.

Chrysanthemum

DIVISION; BASAL CUTTINGS; SEED

Chrysanthemum maximum, Shasta Daisy, has single or double white flowers with a yellow centre. They bloom from summer through to autumn. *Chrysanthemum indicum* hybrids produce single or double flowers of yellow, white, purple, red or brown in autumn.

Division takes place in spring. Lift clumps and divide, replanting 18in/45cm apart.

Basal cuttings of chrysanthemum

Basal cuttings are cut away in spring. Prepare when flowering has finished in autumn by cutting down the stems to 12in/30cm. In late autumn, lift the plants and plant them in a box of loam potting compost. Place in a cold frame for the winter. In early spring, cut away basal shoots about 3in/7.5cm long and plant in half peat potting compost and half sand. Place in a propagator at 60°F/16°C or in a warm greenhouse. Once new growth

appears, harden off and transplant to individual 3in/7.5cm pots. Place in a cold frame. Plant out in late spring or early summer.

Seed of *Chrysanthemum maximum* is sown outdoors in flowering positions from late spring to early summer. Germination takes 4 weeks. Thin out to 12in/30cm. Seed of *Chrysanthemum indicum* hybrids can be sown in late winter. Place in a propagator at 70–75°F/21–24°C. Thin out seedlings when they can be handled easily and harden off. Plant out to flowering positions in late spring or early summer. Later sowings from spring onwards can be made in a frame, planting out in early summer, 18in/45cm apart.

Coreopsis

DIVISION; SEED; BASAL CUTTINGS

Grows to 18in/45cm. *Coreopsis grandiflora*, Tickseed, has bright yellow daisy-like flowers, single or double, some with a contrasting brown centre. They bloom from summer through to autumn. Leaves are lance-shaped and deeply incised.

Division takes place every 3 to 4 years in spring. Lift and divide, planting sections 18in/45cm apart.

Seed is sown under glass in early spring at 55–60°F/13–16°C. Germination takes 28 days. When seedlings can be handled easily, transplant to deep trays of loam potting compost. In early autumn, plant out to flowering positions in a sunny spot, 18in/45cm apart. Outdoor sowings can be made in a seed bed from late spring onwards. Plant out to flowering positions in early autumn.

Basal cuttings are taken in late summer (see page 15). Plant in a mixture of two thirds loam potting compost and one third sand. Place in a cold frame through the winter. Plant out in late spring.

Dahlia

DIVISION; BASAL CUTTINGS; SEED

Grows to 3ft/90cm. *Dahlia* hybrids have seemingly endless forms of flowers — white, cream, orange, yellow, red, lavender or pink in colour, and anemone-

flowered, paeony-flowered, pompom and cactus in shape. They bloom from summer through to autumn.

Division takes place in early spring. Bury overwintered tubers in a shallow box of moist peat and keep at a temperature of 45–50°F/7–10°C. When young shoots are 2in/5cm long, divide the clump so that each piece has a part of the crown with one or more tubers. Dust cut edges with fungicide . Plant each piece in a 5in/12.5cm pot with the tuber 1in/2.5cm below the compost surface. Keep at 60°F/16°C and plant out in late spring or early summer.

Cuttings from stored dahlia tubers

Basal cuttings are taken from tubers planted and stored in late winter or early spring in a box of moist peat. Keep at 60–65°F/16–18°C. Cut away shoots when they are 3in/7.5cm long. Remove the lowest leaves and plant in a mixture of moist peat and sand. Keep at 65°F/18°C. Roots form in about 3 weeks. Transplant to 4in/10cm pots and place in a cold frame. Plant out in late spring when all danger of frost is past. This method produces earlier and probably better flowers.

Seed is sown in late winter. Place in a propagator at 65–70°F/18–21°C. Germination takes 3 weeks. Thin out seedlings when they can be handled easily and harden off. Plant out to flowering positions in late spring, 18in/45cm apart.

Delphinium

DIVISION; BASAL CUTTINGS; SEED

Grows to 4ft/1.2m. *Delphinium* Belladonna Hybrids have tall spikes of loose flowers, blue, pink or white, from summer through to autumn. The foliage is deeply cut. *Delphinium* Large-flowered Hybrids have spikes of more closely-packed flowers in white, blue (various shades) mauve and pink. Some are bicoloured. They bloom in summer.

Division takes place in early spring. Lift and divide large clumps, preferably into only 2 pieces.
Basal cuttings are taken in spring. Take pieces 3–4in/7.5–10cm long, close to the rootstock. Plant in a mixture of two thirds loam potting compost and one third sand. Place in a propagator at 50–55°F/10–13°C or in a cold frame. Harden off and plant out to flowering positions in early autumn.
Seed is sown in late spring in a bed outdoors, covering to a depth of 0.25in/0.5cm. Germination takes 28 days. Thin out seedlings to 6in/15cm when they can be handled easily. Plant out to flowering positions in early autumn, 24in/60cm apart.

Dianthus

LAYERING; CUTTINGS; SEED

Grows to 18in/45cm. *Dianthus caryophyllus*, Border Carnation, was the original from which the many hardy hybrids have developed. Double and single flowers in shades of red, pink, and white are produced during summer months. *Dianthus plumarius*, Pinks, supplied the basis for many hybrids, with flowers of similar shades to the Border Carnations.

Layering of *Dianthus caryophyllus* takes place in summer. Choose shoots close to the soil and, with a sharp knife, make an incision along the stem about 1in/2.5cm long. With a hoop of wire, pin down the stem into a pot of loam potting compost let into the soil. Rooting takes about 4 weeks. Sever from the parent plant when growing well. Overwinter in a cold frame and plant out the following spring.

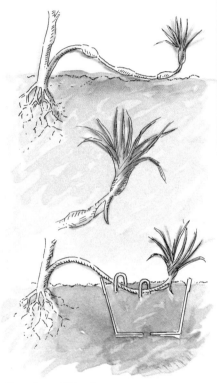

Layering of *Dianthus*

Cuttings of *Dianthus caryophyllus* are taken in summer. Plant cuttings 4in/10cm long in a mixture of peat and sand. Overwinter in a frame and plant out in spring.
With *Dianthus plumarius*, take slips — new shoots with a heel of older wood — or pipings — new shoots pulled out at a socket on the stem — in summer. Remove lower leaves and plant in 3in/7.5cm pots of peat and sand. Overwinter in a frame and plant out the following spring.
Seed is sown in late winter or early spring in trays of loam seed compost. Place in a propagator at 60–70°F/16–21°C. Germination takes 21 days. Thin out

seedlings when they can be handled easily and harden off. Plant out to flowering positions in late spring, 12in/30cm apart.

Echinops

DIVISION; SEED; ROOT CUTTINGS

Grows to 3ft/90cm. *Echinops ritro*, Globe Thistle, has grey-green, deeply-cut leaves, downy underneath, and globular heads of blue flowers in summer.

Division takes place in spring. Lift plants and divide, replanting 2ft/60cm apart.
Seed is sown in a seed bed from spring onwards. Germination takes 6 weeks. Thin out to 12in/30cm when seedlings can be handled easily. In autumn, plant out to flowering positions, 2ft/60cm apart.
Root cuttings are taken in late autumn or early winter. Lift plants and cut off young healthy roots close to the crown of the plant. Remove any dead roots. Make a straight cut across the root where it was joined to the crown, and a slanting cut 2in/5cm further along. Continue making similar straight and slanting cuts every 2in/5cm along the length of the root. Plant vertically in a pot of peat potting compost with the slanting cut end downwards. Leave the straight cut end just showing above the compost surface. Place in a cold frame and plant out to flowering positions in late summer, 2ft/60cm apart.

Erigeron

DIVISION; SEED

Grows to 2ft/60cm. *Erigeron* hybrids, Fleabane, have daisy-like flowers in pink, blue, mauve or violet-mauve. They bloom from summer through to autumn.

Division takes place in spring. Lift and divide, replanting 12–16in/30–40cm apart, depending on the size of the clump.
Seed is sown outdoors in a seed bed from spring onwards. Germination takes 28 days. Thin out seedlings to 6in/15cm when they can be handled easily. Plant out to flowering positions in autumn, 12in/30cm apart. Choose a sunny spot.

Herbaceous Perennials

Gentiana

SEED; DIVISION

Grows to 2ft/60cm. *Gentiana ascelpiadea*, Willow Gentian, has lance-shaped leaves and tubular blue flowers on arching stems in late summer and early autumn. *Gentiana verna*, Spring Gentian, growing to 4in/10cm, produces short, tubular, royal blue flowers flaring out to a 5-petalled star shape. They bloom in spring.

Seed is sown in autumn in a tray of loam seed compost. Leave outside to expose to frost. In spring, bring into the greenhouse or place in a propagator at 60°F/16°C. Germination can be erratic, but is more likely to be successful if seeds have been exposed to frost. Thin out seedlings when they can be handled easily and harden off. Plant out to flowering positions in autumn, 12in/30cm apart.
Division of large clumps takes place in spring or autumn, for spring flowering plants. Replant 12in/30cm apart. Gentians are best left undisturbed, however.

Geum

DIVISION; SEED

Grows to 2ft/60cm. *Geum chiloense*, Avens, has red, yellow, orange, or orange-yellow flowers, similar to a wild rose, in summer. The mid-green leaves are lobed.

Division takes place every 3 to 4 years. Lift plants and divide, replanting 18in/45cm apart in sun or partial shade.

REJUVENATION

Herbaceous perennials send up new shoots in spring after dying down in autumn, but even so, they will not go on for ever. Like all living things they age and lose their vigour, some faster than others. In time they will die. Long before this happens they should be rejuvenated. The miracle is achieved by propagation.

Simple division is the easiest way for most herbaceous perennials. Each year the roots spread outwards, putting up new shoots. As the clump grows, its central part ages, producing weaker shoots and fewer flowers. To help younger growth take over, dig up the plant, preferably in early spring, and remove the older growth. Throw this central part away, and replant sizeable groups of young roots growing at the edges.

Pansies (page 38) are treated as annuals because as perennials they are so short-lived. But they can be rejuvenated by the form of division called Irishman's cuttings, which means teasing out and replanting the plant's loosely growing roots (see page 17).

Rejuvenation can also be achieved by cuttings, particularly with pinks, which, if not renewed, quickly go straggly. They may be given a new lease of life by planting pipings, a form of cutting peculiar to them (see page 17).

Seed is sown in late winter. Place in a propagator at 65–70°F/18–21°C. Germination takes 6 weeks. Thin out seedlings when they can be handled easily and harden off. Plant out to flowering positions in early autumn, 18in/45cm apart. Seed can be sown in a bed outdoors or in a cold frame from late spring onwards, planting out in early autumn.

Helleborus

DIVISION; SEED

Grows to 12in/30cm. *Helleborus niger*, Christmas Rose, has white, saucer-shaped flowers in winter. These later turn green, with yellow anthers. The plant has dark green, tough, leathery leaves.

Division takes place when flowering is over. Lift plants and divide, replanting 12in/30cm apart in a shady spot.
Seed is sown in pots of loam seed compost in autumn. Leave outside in winter, exposed to frost. Germination is probable by spring, but cannot be relied on. It may be necessary to stand seed outside in winter for a second year. Thin out seedlings when they can be handled easily and place in a cold frame. Plant out to a nursery bed in autumn and grow there for a further year before planting out to flowering positions. The plant will not flower until about 3 years old.

Hemerocallis

DIVISION; SEED

Grows to 2.5ft/75cm. *Hemerocallis* hybrids, Day Lily, produces red, pink, orange or yellow lily-like flowers in summer. Some are bicoloured. There are arching, narrow strap-shaped leaves.

Division takes place in spring, but only when clumps become overcrowded. Lift and replant the divided sections, 18in/45cm apart, in sun or partial shade.
Seed is sown in a cold frame from spring onwards. Thin out seedlings when they can be handled easily. Plant out to

flowering positions in autumn, 18in/45cm apart. Flowers bloom within 2 years of sowing.

Hosta

DIVISION; SEED

Grows to 2ft/60cm. *Hosta fortunei*, Plantain Lily, has broad, bluish-green leaves and spikes of lilac flowers in late spring and early summer. Some cultivars have attractive variegated leaves of yellow and green. These plants are grown mainly for foliage.

Division takes place in spring, or leave until autumn in order not to disturb flowering. Lift large clumps and pull or cut away pieces, rather than dividing the whole clump into 4 pieces. Replant, 18in/45cm apart, in a shaded, moist position.
Seed is sown from spring onwards in a tray of loam seed compost. Leave outdoors or place in a cold frame. Germination takes 12 weeks. Thin out seedlings when they can be handled easily. Compost must be moist at all times. Plant out in autumn, 18in/45cm apart.

Kniphofia

DIVISION; SEED

Grows to 4ft/1.2m. *Kniphofia* hybrids, Red Hot Poker, have stout spikes of closely-packed, red, yellow, orange, cream, or bicoloured tubular flowers in summer. Grey-green narrow leaves are sword-shaped. There are also smaller forms, growing to 3ft/90cm.

Division takes place in spring. Lift and divide, replanting 18in/45cm apart in a warm, sunny position.
Seed is sown in late winter in loam seed compost. Place in a propagator at 60–65°F/16–18°C. Germination takes 2 months. Thin out seedlings when they can be handled easily and harden off. Plant out to flowering positions, 18in/45cm apart. Sowings can be made directly into a cold frame from early summer onwards, planting out the following spring.

Lupinus

SEED; SOFT BASAL CUTTINGS

Grows to 2ft/60cm. *Lupinus polyphyllus* hybrids, Lupin, have tapering spikes of closely packed yellow, white, red, blue or bicoloured flowers from late spring to summer. Bright green leaves consist of oval leaflets in circular formation.

Seed is sown in late winter. Soak in tepid water for 24 hours and sow in loam seed compost. Place in a propagator at 55–60°F/13–16°C. Germination takes 6–8 weeks. Thin out seedlings when they can be handled easily and plant out in early summer. Later sowings can be made from spring onwards in a cold frame or greenhouse, or straight into the ground. Plant out to flowering positions in late summer, 18in/45cm apart. Flowers should appear the following year, but early sown seeds may flower the same autumn.
Soft basal cuttings are taken in spring. Use 4in/10cm cuttings with a heel and plant in a mixture of two thirds loam compost and one third sand. Place in a cold frame or propagator at 55°F/13°C. When growing well, plant out in early summer, 18in/45cm apart.

Meconopsis

DIVISION; SEED

Grows to 3ft/90cm. *Meconopsis betonicifolia*, Himalayan Blue Poppy, has sky-blue flowers in early summer. *Meconopsis cambrica*, Welsh Poppy, produces circular yellow to orange flowers from early summer onwards. The plant has deeply incised leaves and self-seeds easily.

Division takes place in early autumn every 3 years. Lift clumps and divide, replanting 18in/45cm apart.
Seed is sown in early winter in a pot of loam seed compost, barely covering the seed. Place outside to expose to frost. In late winter, place in a propagator at 55–60°F/13–16°C. Germination takes 6 weeks. Thin out seedlings when they can be handled easily and harden off. Plant out to flowering positions in early

Meconopsis betonicifolia

autumn, 18in/45cm apart. Seed of *Meconopsis cambrica* can be sown in a bed from late spring onwards and planted out to flowering positions in early autumn.

Paeonia

DIVISION; SEED

Grows to 3ft/90cm. *Paeonia* hybrids, Chinese hybrids, derived in the main from *Paeonia lactiflora*, have single or double flowers in early summer. These are white and pink to deep red. Foliage is dark green.

Division takes place in early autumn. Plants should be divided, at most, every 5 years. Leaving them for a longer period is advisable, because root disturbance may stop them from flowering for a year or two. Lift crowns in early autumn and cut away old unhealthy sections. Divide the remainder of the new vigorous growth, ensuring that each piece has several growing points, or eyes. Replant 2ft/60cm apart, in a sunny or partially shaded spot.
Seed is sown in spring in a cold frame. Germination is erratic and may take up to a year. Thin out seedlings when they can be handled easily. Grow on in the cold frame, planting out to flowering positions in early autumn.

Herbaceous Perennials

Papaver

SEED; ROOT CUTTINGS

Grows to 3ft/90cm. *Papaver orientale*, Oriental Poppy, has large flowers in scarlet, orange, pink or white. They bloom in late spring and early summer. The plant has large, hairy, grey-green, deeply cut leaves.

Seed is sown from early summer in a seed bed. Germination takes 4 weeks. Thin out to 6in/15cm when seedlings can be handled easily. Transplant to flowering positions in early autumn, 18in/45cm apart. Likes full sun.

Rooting cuttings of *Papaver orientale*

Root cuttings are taken in early spring. Lift plants and cut off young healthy roots close to the crown of the plant. Trim away dead roots. Make a straight cut across the root where it joins the crown and a slanting cut 2in/5cm further along. Continue making straight and slanting cuts every 2in/5cm along the length of the root. Plant vertically in a pot of peat potting compost, slanting cut end downwards. Leave the straight cut end just showing above the compost surface. Place in a cold frame and plant out to flowering positions in early autumn, 18in/45cm apart.

Pelargonium

SEED; SOFTWOOD CUTTINGS; SEMI-RIPE CUTTINGS

Pelargonium × hortorum

Grows from 18in/45cm upwards, depending on variety. Strictly a sub-shrub, the familiar species known as geraniums are treated either as half-hardy annuals or tender perennials. The zonal pelargoniums, which have coloured horsehoe-shaped markings on their leaves, are *Pelargonium × hortorum* hybrids. Regal pelargoniums (or Martha Washington pelargoniums) are varieties of *P. × domesticum*, and more tender. The trailing ivy-leaved pelargoniums are *P. peltatum*.

Seed is sown in winter or late winter indoors. F_1 hybrids of zonal pelargoniums are the best choice for raising from seeds, expensive though they are. Sow preferably in peat pots to reduce root disturbance later, and keep near 70°F/21°C. Germination takes 1–2 weeks. Harden off seedlings towards the end of spring, before planting out when all danger of frost is past. They will start to flower in early summer or summer. **Softwood cuttings** are taken in late spring and summer. Take stem-tip cuttings about 4in/10cm long. Remove leaves from lower down the stems, leaving a few small ones at the top. Insert each cutting in a small peat pot. Keep around 65°F/18°C. After the first good watering, water less liberally or the cutting my rot. Rooting takes only a few weeks. Harden off and transplant, still in the peat pot, to a larger pot. Overwinter in a frost-free greenhouse or cold frame.

Taking cuttings of *Pelargonium*

Semi-ripe cuttings are taken in late summer. These may be preferred to softwood cuttings because the stems are less sappy and less liable to rot. Cut off 4–6in/10–15cm long pieces of the current year's woody growth. Trim the cutting to just below a node and remove the lower leaves, but not the tip. Dip in hormone rooting powder and plant in a peat pot. Keep at 65°F/18°C until rooted, with signs of new growth. Move into slightly larger pots and gradually harden off to overwinter in a cold frame, adequately covered to keep out frost when necessary. Plant out from spring onwards, when all danger of frost has passed.

Primula

DIVISION; SEED

Grows to 12in/30cm. *Primula denticulata*, Drumstick Primrose, has large oval leaves and globular heads of mauve flowers on stout stems during spring months. *Primula vulgaris*, Common Primrose, grows to 6in/15cm, producing bright yellow flowers with a darker shaded eye in spring.

Division takes place when flowering is over. Lift clumps and divide, replanting 12in/30cm apart, in a sunny or slightly shaded spot.
Seed is sown in early winter. Rinse the seed in cold water and plant in pots of loam seed compost, just covering with compost. Place outside to expose to frost. In early spring, place in a propagator at 55–60°F/13–16°C. Germination takes 6 weeks. Thin out seedlings when easy to harden off. Transfer to a frame and plant out to flowering positions in early autumn.

Trollius

DIVISION; SEED

Grows to 2.5ft/75cm. *Trollius × cultorum*, Globe Flower, has globe-shaped yellow or orange flowers in spring, with dark green, deeply incised leaves.

Division takes place in early autumn.

Lift clumps and divide, replanting 18in/45cm apart, in full sun or partial shade. Plants increase slowly and are best left for several years before dividing.
Seed is sown in winter in a pot of loam seed compost. Place outdoors to stratify. In early spring, place in a propagator at 55–60°F/13–16°C. When seedlings can be easily handled, thin out and harden off. Place in a cold frame and plant out to flowering positions in autumn. Seed may take some time to germinate; if so, wait to plant them out in autumn the following year.

Verbascum

SEED; DIVISION; ROOT CUTTINGS

Grows to 3ft/90cm. *Verbascum phoeniceum* hybrids, Mullein, have spikes of white, pink, purple, yellow or blue saucer-shaped flowers from late spring to summer.

Seed is sown from spring onwards in a seed bed. Germination takes 6 weeks. Thin out seedlings when they can be handled easily. Transplant to flowering positions in early autumn, 18in/45cm apart in a sunny spot.
Division takes place in autumn after flowering. Lift plants and divide, replanting 18in/45cm apart.
Root cuttings are taken in late autumn or early winter. Lift plants and cut off young healthy roots close to the crown of the plant. Trim off any dead roots. Make a

straight cut across the root where it joined the crown and a slanting cut 2in/5cm further along. Continue making similar straight and slanting cuts every 2in/5cm along the length of the root. Plant vertically in a pot of peat potting compost with the slanting cut end downwards. Leave the straight cut end just showing above the compost surface. Place in a cold frame and plant out to flowering positions in early autumn, 18in/45cm apart.

Zantedeschia

DIVISION OF RHIZOME; SEED

Grows to 3ft/90cm. *Zantedeschia aethiopica* 'Crowborough', Arum Lily, is a hardy form, producing pure white spathes surrounding yellow spadices in summer. Its bright green leaves are spear-shaped.

Division of rhizome takes place in spring. The rhizomes, which increase readily, are lifted and divided, planting 18in/45cm apart. Make sure each piece has plenty of growing points.
Seed is sown in spring. Soak the seed in warm water for 24 hours and sow in loam seed compost, covering to the depth of the seed. Place in a propagator at 70–75°F/21–24°C. Germination may take up to 12 weeks. Thin out seedlings when they can be handled easily and harden off. Place in a cold frame and grow on there before planting out in spring the following year. Needs protection in very cold, frosty areas; grow in pots in the greenhouse, placing them outside in summer.

THE NURSERY BED

A nursery bed is the temporary home for a plant between the seedling stage and adolescence, before it moves to its permanent home. For hardier plants it can be an alternative to the protection of a cold frame. All you need is a little patch of garden set aside just for plants to grow up in. Nothing elaborate is required as long as the ground is well-drained and fertile, not too exposed to the sun but not in deep shade, and, above all, sheltered from chilling winds. By concentrating young plants in one part

of the garden you can give them more constant attention, especially in watering and weeding, than if they were scattered here and there.

To ease the duty of after-care, keep the young plants regimented in rows or blocks, as in the seed bed, and clearly label, perhaps adding the dates of sowing and transplanting.

When planting in the nursery bed, allow spacing of 6–10in/15–25cm apart, according to size, using a large dibber or trowel to make holes big enough to avoid cramping the roots.

Shrubs

With the changing fashions in gardening, the 'carpet bedding' of annuals was succeeded at the end of the 19th century by borders of herbaceous perennials, until these were in turn superseded by shrub borders in the mid-20th century. So in the space of a hundred years, gardeners transferred their preference from short-lived to long-lived plants, and progressively from labour-intensive gardening to labour-saving. Now it is the 'mixed border' which dominates the garden, but shrubs remain the 'backbone', as the cliché goes.

There is much to be said for shrubs. Even in winter they demonstrate advantages over the herbaceous perennials. Evergreen shrubs are in leaf, while the bare stems of deciduous shrubs reveal distinctive silhouettes hidden in summer. Some retain the berries of autumn; others have the swelling buds of spring. One or two are even in flower. From early spring to spring, more and more shrubs will come into bloom, with the great burst to follow in late spring and early summer. The abundance tails off from late summer, but in a mixed border, herbaceous perennials and autumn bulbs can fill the gaps.

Avoid overcrowding shrubs, as in a Victorian shrubbery, or in the contemporary name of 'ground cover'. The temptation to pack plants closely together is all the greater if you have raised them yourself, forgetting that the successfully germinated seedlings or rooted cuttings may, in time, be as tall as you are, and probably wider.

NEW SHRUBS

Shrubs can be propagated from seeds, by cuttings (rarely softwood, more often semi-ripe wood, but mainly hardwood), by simple layering, and a few by

Most shrubs are excellent subjects for a variety of propagation methods. But beware of spoiling the appearance of a favourite plant by too eager collection of cuttings or ruthless division. It may take some time for older shrubs to renew their pleasing shape, or for young cuttings, such as these of *Syringa vulgaris*, Lilac, and *Senecio laxifolius*, left, to develop into mature plants.

Shrubs

serpentine layering and leaf bud cuttings.

Seed of most shrubs which survive outdoors can be sown outdoors in seed beds, although some will do better if started indoors with a little heat.

When sowing outdoors, dig the site for the seed bed in autumn; 36in/90cm is wide enough and the length can be variable. In early spring or spring, weather permitting, rake the bed level and sow the seed in shallow drills 4–6in/10–15cm apart. Fine seed needs very little covering; larger seeds should be 0.25–0.5in/0.5–1.25cm deep. Thin out the seedlings as needed. Water the bed in dry weather and keep it weeded. The young plants stay in the bed until autumn or spring of the following year, when they are moved to a nursery bed or the garden, depending on how robust they are.

Seeds which are better sown indoors include those of camellia, cytisus, erica, mahonia, potentilla, rhododendron and viburnum. Sow in seed trays or pots in a loam seed compost; a lime free compost should be used for camellias, ericas, pernettyas and rhododendrons. There is no need to cover small seeds, but large seeds require a shallow covering of sieved compost. Water by putting the tray/pot into a basin of shallow, tepid water, leaving it until moisture works its way to the surface. Then let it drain. Cover with a piece of glass and a sheet of newspaper and place in a propagator. Wipe off drops of moisture from the underside of the glass each day, and as soon as the seeds start to germinate, remove the newspaper.

Prick out the seedlings into loam potting compost. Keep in the propagator for a little longer, gradually hardening them off to go into a cold frame. If they look sturdy enough to stand the winter, remove them to a sheltered nursery bed in the autumn. Otherwise keep them in the frame over winter and transplant in spring. When they are large enough, move them from the nursery bed to their place in the garden. Do this from late autumn to early spring for deciduous shrubs and trees. Move evergreens either from early autumn to autumn, or leave them until early spring to spring.

Semi-ripe cuttings, from the woody part of the current year's growth, are taken from summer to autumn. Those taken from shrubs are commonly 4–6in/10–15cm long. The still-soft tip of the cutting is removed. The other end is trimmed to just below a node, dipped in hormone rooting powder and the lower leaves removed.

The semi-ripe cuttings of many shrubs can be rooted in a cold frame. They may be planted directly in

Semi-ripe cuttings

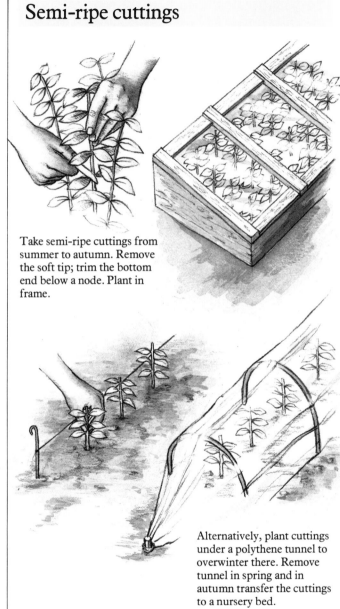

Take semi-ripe cuttings from summer to autumn. Remove the soft tip; trim the bottom end below a node. Plant in frame.

Alternatively, plant cuttings under a polythene tunnel to overwinter there. Remove tunnel in spring and in autumn transfer the cuttings to a nursery bed.

the frame, or in pots or boxes which are then placed in the frame. When planting in pots, use a seed compost or a mixture of peat and sand and insert the cutting to about a quarter of its length. If planting directly in the frame, cover the soil with a 3–4in/7.5–10cm layer of compost and plant the cuttings some 3in/7.5cm apart. Let them spend the winter there with the lights firmly closed. They may not root until spring and will not be ready to move to the nursery bed until autumn.

Hardwood cuttings

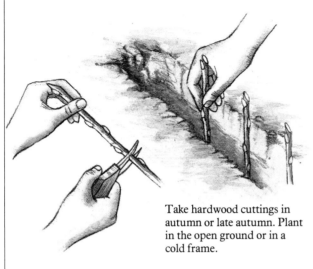

Take hardwood cuttings in autumn or late autumn. Plant in the open ground or in a cold frame.

Heel cutting

Heel cuttings are semi-ripe or hardwood cuttings with a 'heel' of bark from the main stem. Trim before planting.

An alternative to the cold frame is to plant the cuttings under a polythene tunnel, 36in/90cm wide. Put a layer of compost, 3in/7.5cm deep, on top of the soil where the cuttings are to be planted. Place them in rows 4in/10cm apart, with 3in/7.5cm between cuttings. In the following spring, remove the polythene. By the autumn, well-established plants should be ready to move to the nursery bed.

A variation of the semi-ripe cutting is a **mallet**

Simple layering

A branch near the ground can be rooted by making a nick in the stem before partly burying it in the soil.

cutting — which describes its appearance. Berberis with thin stems are often treated in this way, since thin stems often root less readily. The cutting consists of a current year's shoot attached to a short piece of old wood, 0.25–0.5in/0.5–1.25cm long, from which roots are more likely to grow.

HARDWOOD CUTTINGS

Hardwood cuttings are likely to have a higher success rate than semi-ripe cuttings, and far higher than softwood cuttings. Hardwood cuttings of deciduous shrubs are usually taken in autumn or late autumn, after the leaves have fallen. Remove cuttings 8–10in/20–25cm long, making a slanting cut just above a bud and a straight cut just below a node. It is the straight end which goes into the soil. Dip it in hormone rooting powder.

If planting in the open ground, dig a narrow, wedge-shaped trench; one side should be vertical and the other slanting. This can often be achieved simply by pushing a spade straight down into the soil; then lever it away from you to create a wide enough gap. Insert the cuttings against the vertical side of the trench, 4–6in/10–15cm apart, and firm the soil around them. If during the winter months frost lifts the

Shrubs

cuttings out of the soil, firm them in again as soon as the frost goes. Leave them until the autumn and then move to the nursery bed.

Most shrub cuttings will do better during the first winter if given the protection of a cold frame, however. See that the lights are closed. From spring or late spring onwards, the lights can be opened more and more, and finally removed. The cuttings are left in the frame throughout the summer months and by autumn, many will be large enough to move to permanent positions. Any laggards can be moved to the nursery bed for a year.

Heel cuttings are hardwood or semi-ripe side shoots pulled from the main stem with a 'heel' of its bark. Any jagged ends are trimmed from the 'heel' to lessen the risk of rotting; otherwise the cuttings are treated as semi-ripe or hardwood cuttings.

LAYERING

Simple layering is used to propagate a number of shrubs with low branches, which will grow roots if buried in the ground. Plants can do it for themselves, but humans have been helping them since ancient times; indeed, the Latin verb *propagare*, from which 'propagation' is derived, means multiplying by layers. The great merit of layering is that the piece of stem buried in the ground remains attached to, and sustained by, the parent plant until it has roots enough to live on its own. The time which this takes varies from shrub to shrub. Some may take only six months; some a year; and rhododendrons and mahonias up to two years.

Layering is best carried out between spring and late summer, when the plant is active and the soil warm. Choose a stem which has grown the same year or the year before, rather than an old one which takes longer to root. First wound the stem about 12in/30cm away from the tip, by making a cut at an angle halfway through the stem. Wedge it open with a matchstick. Then in the ground beneath the stem, dig a slanting hole, 4–6in/10–15cm deep, so that its centre is at the point where the wound on the stem will be buried. Remove any leaves that would be covered by soil. Peg the stem in the hole with a bent piece of galvanized wire and tie the tip with soft string or raffia to a cane, to make it grow erect. Pile 6in/15cm of soil or compost over the stem in the hole. Keep the layer moist during the summer months and leave it attached to the parent plant until the following spring. Then it can be severed and soon afterwards transplanted, if it has grown plenty of roots. If not, allow it more time.

Leaf bud cuttings can be used to grow new plants of camellia and mahonia (see the individual entries for these plants).

Berberis

SEMI-RIPE CUTTINGS; SEED; DIVISION

Grows to 6ft/1.8m. *Berberis thunbergii*, Barbery, produces yellow flowers from late spring to early summer, followed by scarlet berries. The variety 'Atropurpurea' has reddish-purple leaves and 'Atropurpurea Nana' is a dwarf form, growing to about 2ft/60cm.

Semi-ripe cuttings are taken in late summer. Use 4in/10cm cuttings with a heel or, preferably, with a short section of old wood — about 0.5in/1.25cm long — a mallet cutting. Dip cuttings in hormone rooting powder and plant in a cold frame in a mixture of half peat and half sand, or in a propagator with gentle bottom heat of 55–60°F/13–16°C. When growing well, gradually harden off cuttings raised in a propagator and place in a cold frame.

Overwinter in the cold frame and plant out the following autumn.

Seed is sown in a seed bed in autumn to stratify. Germination takes place during spring. Thin out as necessary. Plant out in the autumn or wait until the following spring if growth is weak. Alternatively, stratify seed by placing it outdoors in a pot of sand during the winter. Then sow in loam seed compost in early spring and place in a propagator at 55–60°F/13–16°C. When growing well, harden off and transfer to a cold frame. Plant out in the autumn or the following spring.

Division takes place in winter. Some species produce suckers which can be detached. Lift plants, divide and replant immediately.

Buddleia

SOFTWOOD CUTTINGS;
HARDWOOD CUTTINGS; SEED

Grows to 10ft/3m. *Buddleia davidii*,

Butterfly Bush, produces long, tapering spikes of closely packed purple, reddish-purple or white flowers from summer through to autumn. Grows quickly and rampantly.

Softwood cuttings are taken in spring. Use 4in/10cm cuttings with a heel. Plant in a mixture of half peat and half sand and place in a propagator at 55–60°F/13–16°C. When growing well, harden off and transfer to a cold frame. Plant out in autumn the following year.

Hardwood cuttings are taken in late autumn, and are likely to be more successful than softwood cuttings. Use pieces 10in/25cm long, making a slanting cut just above a bud and a straight cut just below a node. Dip the straight cut end in hormone rooting powder and plant in a cold frame. Plant out the following autumn.

Seed is sown in spring, in shallow drills in a prepared seed bed. Cover the fine seed with a thin layer of soil. Thin out as necessary and plant out in autumn. Wait until spring the following year if plants look weak.

Camellia

SEMI-RIPE CUTTINGS; LEAF BUD CUTTINGS; LAYERING; SEED

Old established plants can grow to 20ft/6m or more. *Camellia japonica* includes a large number of varieties, producing single, semi-double and double cup-shaped red, pink, white and bi-coloured flowers from late winter to spring. The evergreen leaves are glossy and leathery.

Semi-ripe cuttings are taken in late summer. Use pieces 4in/10cm long and plant in a mixture of half peat and half sand. Place in a propagator at 60°F/16°C. When well established, harden off and transfer to a cold frame. Overwinter there and plant out in autumn the following year.

Leaf bud cuttings are taken in late summer. Cut a shoot about 10in/25cm long from the present year's growth. Cut into pieces, each with a leaf, making the top cut just above a leaf and the bottom cut 1in/2.5cm below. Plant in a mixture of half peat and half sand and place in a propagator at 60°F/16°C. Thereafter

treat as semi-ripe cuttings.

Layering takes place in late summer. Select a shoot from the present year's growth. Make a cut at an angle half-way through the stem, about 12in/30cm from the tip. Wedge open with a matchstick. Dig a hole and peg down the cut section. Cover with soil and support the tip with a cane to make it grow upright. Sever from the parent the following spring and transplant if there are plenty of roots. Otherwise leave undisturbed for a little longer.

Seed is sown in spring. Soak the seed in warm water for 24 hours as a further help to germination. File the seed coat so that it can absorb moisture — a safer method than cutting it with a knife. Sow in loam seed compost, covering to a depth of 0.25in/6mm. Place in a propagator at 65°F/18°C. When growing well, harden off and transfer to a cold frame or greenhouse. Grow there until autumn the following year. Plant out in a sheltered spot in lime-free soil.

Ceanothus

SEMI-RIPE CUTTINGS; SEED

Grows to 10ft/3m. *Ceanothus* hybrids, Californian Lilac, produce rich, deep blue to mauve flowers in spring or autumn, sometimes both, depending on the hybrid. There are evergreen and deciduous forms.

Semi-ripe cuttings are taken in late summer. Use a 4in/10cm cutting with a

heel and plant in a mixture of half peat and half sand. Place in a propagator at 65°F/18°C. Once well established, harden off and transfer to a garden frame. Grow on there, planting out in autumn the following year.

Seed is sown in spring. Soak seed for 24 hours in warm water. Sow thinly in loam seed compost and place in a propagator at 60–65°F/16–18°C. When growing well, harden off and transfer to a frame. Overwinter there and plant out in autumn the following year.

Chaenomeles

LAYERING; HARDWOOD CUTTINGS; SEED

Grows to 6ft/1.8m. *Chaenomeles speciosa*, Ornamental Quince, produces deep red, pink or white flowers in spring, depending on the variety, followed by yellow fruits in summer. Prune back new shoots in late autumn as this shrub flowers on old wood.

Layering takes place in summer. Select a shoot from the present year's growth. Make a cut at an angle halfway through the stem, about 12in/30cm from the tip. Wedge open with a matchstick. Dig a hole and peg down the cut section. Cover with soil and support the tip with a cane to make it grow upright. Sever from the parent the following spring and transplant, if there are plenty of roots. Otherwise leave undisturbed for a little longer.

Hardwood cuttings are taken in late autumn. Use 10in/25cm pieces, making a slanting cut just above a bud and a straight cut just below a node. Dip the straight end in hormone rooting powder and plant in a cold frame. Plant out the following autumn.

Seed is sown in late winter in a cold frame. Germination takes about 6 weeks. Thin out seedlings when they can be handled easily, transplanting to 3in/7.5cm pots of loam potting compost. Grow on in a frame or greenhouse, planting out in autumn the following year. It may take up to 5 years to reach flowering size.

Shrubs

Cotoneaster

SEMI-RIPE CUTTINGS; LAYERING SEED

Grows to 5ft/1.5m. *Cotoneaster horizontalis* has small, shiny leaves arranged along the branches in herringbone pattern. These turn pink to scarlet in late autumn and early winter. Bright red berries are produced in autumn.

Semi-ripe cuttings are taken in late summer. Use pieces 4in/10cm long and plant in a mixture of half peat and half sand. Place in a propagator at 65°F/18°C. Once well established, harden off and transfer to a garden frame, planting out in autumn the following year.
Layering takes place in late summer. Choose a shoot from the present year's growth and make a cut at an angle halfway through the stem, 12in/30cm from the tip. Wedge open with a matchstick. Dig a hole and peg down the cut section. Cover with soil and support the tip with a cane to make it grow upright. Sever from the parent the following spring. Transplant if there are plenty of roots, otherwise leave undisturbed for a little longer.
Seed is sown in late autumn in a seed bed, or is placed in a pot of sand to stratify during the winter. In late winter, vigorously rub the seed and some sand together between your hands to chip the hard outer coating. Sow in a seed bed, just covering with soil. Thin out, if necessary. Plant out in the autumn or, for less robust plants, in spring the following year.

Cytisus

SEMI-RIPE CUTTINGS; SEED

Grows to 10ft/3m. *Cytisus battandieri*, Broom, has silvery-grey leaves divided into 3 leaflets, and yellow pineapple-scented flowers in summer. *Cytisus scoparius*, Yellow Broom, growing to 6ft/1.8m, has varieties producing yellow and red flowers and combinations of the colours in summer.

Semi-ripe cuttings are taken in late summer. Use 4in/10cm cuttings with a heel and plant in a mixture of half peat and half sand. Place in a propagator at

60–65°F/16–18°C. When well established, harden off and place in a cold frame. Overwinter there and plant out in autumn the following year.
Seed is sown in early spring. Soak for 24 hours before sowing in loam seed compost. Place in a propagator at 65°F/18°C. Germination takes about 8 weeks. When growing well, harden off and transfer to a cold frame. Grow on there and plant out in autumn of the following year. Flowers bloom 2 or 3 years after sowing.

Daphne

SEMI-RIPE CUTTINGS; SEED; LAYERING

Grows to 3ft/90cm. *Daphne mezereum*, Mezereon, has pink, dark red or white flowers in early spring, followed in autumn by red berries on red forms and yellow berries on white forms.

Semi-ripe cuttings are taken in late summer. Use 4in/10cm cuttings and plant in a mixture of half peat and half sand. Place in a propagator at 65°F/18°C. Once well established, gradually harden off and transfer to a garden frame. Overwinter there and plant out in autumn the following year.

Seed is stratified in early winter. Place seed in a pot of sand and stand outside to expose to frost. In early spring, sow in a pot of loam seed compost and place in a propagator at 60°F/16°C. Harden off and transfer to a garden frame. Either transplant to a nursery bed in autumn or leave until the following spring. Plant out to final position in late autumn the following year.
Layering takes place in spring. Select a branch of the previous year's growth and make a slanting cut halfway through the stem, 12in/30cm from the tip. Wedge open with a matchstick. Pin down the cut section, covering with soil. Support the tip of the branch with a cane to make it grow upright. Sever from the parent plant the following spring and transplant to its final position in the garden.

Deutzia

HARDWOOD CUTTINGS; SEMI-RIPE CUTTINGS

Grows to 6ft/1.8m. In early summer, *Deutzia* x *rosea* has clusters of white flowers in the form 'Campanulata', and rose-pink flowers in the form 'Carminea'.

Hardwood cuttings are taken in late autumn and early winter. Use 10in/25cm

---

SCENTED SHRUBS

Mahonia. Lily of the valley scent, stronger in *Mahonia japonica* than in *Mahonia aquifolium*. Flowers in early spring.
Daphne mezereum. Spicy perfume. Flowers in early spring.
Berberis. Delicate sweet perfume from late spring to early summer.
Viburnum carlesii. Almond fragrance in spring.
Rhododendron. Many species and hybrids (such as Azalea, illustrated right) have a rich, spicy perfume. Flowers bloom from early spring to late summer.
Lavandula spica. Inimitable lavender scent in summer.
Cytisus battandieri. Pineapple aroma in summer.
Philadelphus hybrids. Orangey

fragrance. Blooms in summer.
Rosa. Many, but not all, roses have a wide range of scents from the almost imperceptible perfume of tea roses to a penetrating spiciness. Summer blooms.

cuttings with a slanting cut just above a bud and a straight cut just below a node. Dip the straight end in hormone rooting powder and plant in a cold frame. Plant out the following autumn.

Semi-ripe cuttings are taken in summer. Use 4in/10cm stem cuttings. Plant in a mixture of half peat and half sand and place in a cold frame. Overwinter and transfer to a nursery bed in autumn the following year.

Erica

SEMI-RIPE CUTTINGS; SEED

Grows to 1ft/30cm. *Erica carnea*, Winter Heath, produces pink, rose-red, and white flowers from winter through to early spring. *Erica vagans*, Cornish Heath, has varieties with similar flower colour to *Erica carnea*, but produces them from summer to autumn.

Semi-ripe cuttings are taken in summer. Use cuttings 1in/2.5cm long and plant in boxes or shallow pots of half peat and half sand. Place in a frame or greenhouse. Rooting takes about 6 weeks. Transplant to pots of loam potting compost in early spring and plant out to flowering positions in early summer.
Seed is sown in spring, in loam seed compost. Leave uncovered so that light can reach the seed. Place in a propagator at 60–65°F/16–18°C. Germination takes about 12 weeks. Thin out seedlings when they can be handled easily, and harden off. Transfer to a garden frame and overwinter there. Transplant to a nursery bed in spring the following year and plant out to flowering positions in autumn.

Forsythia

SEMI-RIPE CUTTINGS;
HARDWOOD CUTTINGS; LAYERING

Grows to 6ft/1.8m. In spring, *Forsythia* x *intermedia*, Common Forsythia, produces a mass of deep yellow, tubular flowers, flaring to a star shape. The best form is probably 'Spectabilis'.

Semi-ripe cuttings are taken in early summer. Use 4–6in/10–15cm cuttings and plant in a mixture of half peat and

half sand. Place in a propagator at 65°F/18°C. When growing well, harden off and transfer to a garden frame. Grow on over winter and transplant to a nursery bed in autumn the following year.
Hardwood cuttings are taken in late autumn and early winter. Select 10in/25cm cuttings and make a slanting cut just above a bud, and a straight cut just below a node. Dip the straight cut end in hormone rooting powder and plant in a cold frame. Plant out the following autumn.
Layering takes place in summer. Choose a shoot of the present year's growth, making a cut at an angle halfway through the stem, about 12in/30cm from the tip. Wedge open with a matchstick. Pin down the cut section and cover with soil. Support the tip with a cane to make it grow upright. Sever from the parent the following spring and transplant if there is good root structure. Otherwise, leave undisturbed for a little longer.

Garrya

SEMI-RIPE CUTTINGS

Grows to 10ft/3m. *Garrya elliptica*, Silk Tassel Bush, is an evergreen with leaves of dark green and grey undersides. Male

plants have groups of long, greeny-grey catkins in late winter.
Semi-ripe cuttings are taken in late summer. Plant 4in/10cm cuttings in a mixture of half peat and half sand. Place in a propagator at 60°F/16°C. When growing well, transfer to a cold frame. Overwinter there and plant out in autumn the following year in a sheltered part of the garden.

Hebe

SEMI-RIPE CUTTINGS; SOFTWOOD CUTTINGS

Grows to 4ft/1.2m. *Hebe* 'Autumn Glory' is a hybrid form with leathery, oval to round leaves. Clusters of violet flowers are produced from summer through to autumn.

Semi-ripe cuttings are taken from summer to early autumn. Use 4in/10cm cuttings and plant in a cold frame. Leave over winter and plant out in autumn the following year.
Softwood cuttings are taken in late spring. Use 4in/10cm cuttings and plant in a mixture of half peat and half sand. Place in a propagator at 65°F/18°C. When new growth appears, harden off and transfer to a cold frame. Treat then as semi-ripe cuttings.

Shrubs

Hydrangea

SEMI-RIPE CUTTINGS; SEED

Grows to 6ft/1.8m. In summer,
Hydrangea macrophylla, Common
Hydrangea, produces large globular heads
of pink to red or blue flowers.

Semi-ripe cuttings are taken in late
summer. Use 4in/10cm cuttings, from
non-flowering shoots. Plant in a mixture of
half peat and half sand and place in a
propagator at 65°F/18°C. Harden off and
transfer to a cold frame. Overwinter there
and plant out the following autumn.
Seed is sown in spring in loam seed
compost. Place in a propagator at
60–65°F/16–18°C. Thin out and
gradually harden off. Transfer to a cold
frame to overwinter. Plant out the
following autumn.

Hypericum

SEMI-RIPE CUTTINGS; SEED; DIVISION

Grows to 12in/30cm. *Hypericum
calycinum*, Rose of Sharon, is an evergreen
shrub with leathery, dark green, oval
leaves. Bright yellow flowers with darker
coloured stamens appear in summer.
Useful for ground cover, this plant can be
invasive if not regularly pruned.

Semi-ripe cuttings are taken in early
autumn. Use 4in/10cm stem cuttings and
plant in a cold frame. Overwinter there
and plant out to flowering positions the
following autumn. Cuttings can also be
rooted in a propagator at 60°F/16°C,
hardened off and transferred to a cold
frame.
Seed is sown in spring in loam seed
compost. Place in a propagator at
60°F/16°C. Thin out seedlings and
harden off, transferring to a cold frame.
Grow on there and plant out in autumn
the following year.
Division takes place in spring. Lift
dormant plants and divide, replanting
immediately 2ft/60cm apart.

Ilex

SEMI-RIPE CUTTINGS; LAYERING; SEED

Grows to 20ft/6m, but many varieties
reach 6ft/1.8m only. *Ilex aquifolium*,
Common Holly, varieties have shiny,
leathery, dark green leaves, some with
stout spines, others almost spineless.
There are variegated forms with silver or
yellow margins. Scarlet berries appear in
winter.

Semi-ripe cuttings are taken in late
summer. Use 4in/10cm cuttings and dip
in hormone rooting powder. Plant in a
mixture of half peat and half sand. Place
in a propagator at 65°F/18°C. Rooting
can take several months. When cuttings
are growing well, harden off and place in a
cold frame. Plant out in autumn the
following year or wait until spring.
Layering takes place in summer. Choose
a shoot from the present year's growth.
Make a cut at an angle halfway through
the stem, about 12in/30cm from the tip.
Wedge open with a matchstick. Peg down
the cut section and cover with soil. Stake
the tip with a cane to make it grow
upright. Sever from the parent the
following spring and transplant if there is
a good root structure. Otherwise leave for
a while.
Seed is sown in a prepared seed bed in
spring. Thin out as necessary and plant
out in autumn or the following spring.
This method is not recommended,
however, as hollies are slow to grow and
the seed requires stratifying in dry sand
for at least a year.

Lavandula

SEMI-RIPE CUTTINGS; SEED

Grows to 4ft/1.2m, but more compact
forms reach only 2ft/60cm. *Lavandula
spica*, English Lavender, has spikes of
closely packed, tubular, fragrant, mauve
flowers in summer. Some forms produce
white or pink flowers.

Semi-ripe cuttings are taken in late
summer. Use cuttings with a heel and
plant in a cold frame. Overwinter there
and plant out in autumn the following
year.
Seed is sown in spring, in drills in a
prepared seed bed. Germination takes 8
weeks. Thin out seedlings when they can
be handled easily. Transfer to flowering
positions in autumn, 12in/30cm apart.

Magnolia

LAYERING

Grows between 6–12ft/1.8–3.6m.
Magnolia denudata, Lily Tree, is
deciduous, producing white, cup-shaped
flowers in spring.

Layering takes place in late spring and is
probably the best method of propagation.
Plants grown from seed take several years
before they reach flowering size. Choose a
shoot of the previous year's growth and
make a cut at an angle halfway through
the stem, about 12in/30cm from the tip.
Wedge open with a matchstick. Peg down
the cut section in a ready-prepared hole
and cover with soil. Tie the tip to a cane to
make it grow upright. Sever from the
parent plant the following spring and
transplant if it has a good root structure.
Otherwise, leave undisturbed until later
in the year.

Mahonia

LEAF BUD CUTTINGS; SEED; DIVISION

Grows to 3ft/90cm. *Mahonia aquifolium*,
Oregon Grape, has dark green, glossy,
toothed, oval leaves. Dense clusters of
yellow flowers bloom in early spring.

Leaf bud cuttings are taken in autumn.

Cut a shoot about 10in/25cm long from the present year's growth; the wood should be green, not brown. Cut this shoot into pieces, 1in/2.5cm above a pair of leaves and 2in/5cm below. Remove one of the leaves. Plant in a mixture of half peat and half sand, with the base of the leaf just above the surface of the compost. Place in a propagator at 60°F/16°C. Cuttings may take some time to root. When growing well, harden off and transfer to a cold frame. Plant out in autumn the following year.

Seed is stratified in early winter. Bury the seed in a pot of sand and place outside. In early spring, sow the seed in loam seed compost and place in a propagator at 60°F/16°C. Thin out and, when growing well, gradually harden off. Transfer to a cold frame. Plant out in autumn or wait until the following spring if plants are not robust.

Division takes place in autumn. Mahonias increase by the production of underground suckers, and so are self-layering. Lift overcrowded clumps and divide, replanting 2ft/60cm apart.

Pernettya

SEED; SEMI-RIPE CUTTINGS; LAYERING

Grows to 5ft/1.5cm, but there are smaller forms. *Pernettya mucronata*, Prickly Heath, has evergreen leaves and white flowers in late spring, followed by red berries. There are varieties with white, pink, or purple berries. (For a good show of berries a male shrub is usually planted with several females.) Not for limey soils.

Seed is sown in a lime-free peat seed compost in spring. Place in a propagator at 60°F/16°C. When young plants are well established, harden off and transfer to a cold frame. Plant out in autumn or the following spring.

Semi-ripe cuttings of 4in/10cm long are taken in summer or late summer. Dip cuttings in hormone rooting powder and plant in a mixture of half peat and half sand. Place in a propagator at a gentle heat of 55–60°F/13–16°C, or in a cold frame. Plant out the following autumn.

Layering requires no effort from the gardener because these shrubs layer themselves very easily. In autumn, lift healthy, sturdy shoots which have taken root. Sever from the parent plant and replant immediately in the garden where they are destined to grow.

Philadelphus

SEMI-RIPE CUTTINGS; HARDWOOD CUTTINGS; SEED

Grows 4–8ft/1.2–2.4m depending on variety. *Philadelphus* hybrids, Mock Orange, have fragrant flowers, single, semi-double or double. These are white, often with contrasting colour, and produced in summer.

Semi-ripe cuttings are taken in summer. Use 4in/10cm cuttings and plant in a cold frame. Overwinter there and plant out in autumn the following year.

Hardwood cuttings are taken in late autumn and early winter. Use 10in/25cm cuttings, making a slanting cut just above a bud and a straight cut just below a node. Dip the straight cut end in hormone rooting powder and plant in a wedge-shaped trench made in a cold frame. Half to two thirds of the cuttings should be below ground. Plant out in autumn the following year.

Seed is sown in spring in loam seed compost. Place in a propagator at 60°F/16°C. Thin out and harden off when seedlings are growing well. Transfer to a cold frame. Overwinter there, transplanting to a nursery bed in spring. Plant out to final positions in autumn.

Potentilla

SEMI-RIPE CUTTINGS; SEED

Grows to 4ft/1.2m. *Potentilla fructicosa*, Cinquefoil, produces successions of 5-petalled flowers — yellow, copper-orange, or cream — from late spring to early autumn, depending on the hybrid chosen. There are small oval leaves.

Semi-ripe cuttings are taken in late summer. Use 4in/10cm cuttings, removing the soft growing tip. Dip in hormone rooting powder and plant in a cold frame. Overwinter there and move to a nursery bed in autumn the following year.

Seed is sown in spring, in loam seed compost. Place in a propagator at 60–65°F/16–18°C. Thin out and harden off when growing well. Transfer to a cold frame and overwinter there. Transplant to a nursery bed in spring and plant out to final positions in autumn.

EARLY BLOOMS

Winter-early spring *Camellia japonica* (red, pink, white); *Erica carnea* (pink, rose, white).

Early spring *Daphne mezereum (pink, red, white); Mahonia aquifolium* (yellow); Azaleas (many colours).

Spring *Chaenomeles speciosa* (pink, red, white); *Forsythia intermedia* (yellow); *Magnolia denudata* (white); *Rhododendron* (red and many other colours); *Viburnum carlesii* (pink, turning white).

Late spring *Prunus tenella* (bright pink); *Pyracanthus coccinea* (white).

Camellia japonica

Shrubs

Prunus

DIVISION

Shrubs grow to 4ft/1.2m but often less. *Prunus tenella*, Dwarf Russian Almond, has bright pink flowers in late spring, spreads freely from suckers, and has small shiny leaves.

Division takes place in autumn. Lift plants and divide, replanting immediately 2–3ft/60–90cm apart.

Pyracantha

SEMI-RIPE CUTTINGS; SEED

Grows to 6ft/1.8m when treated as a shrub. *Pyracantha coccinea*, Firethorn, produces clusters of small white flowers, similar to hawthorn, in late spring and early summer. Bright red or orange-red berries follow. The evergreen leaves are small, oval and tapering.

Semi-ripe cuttings are taken in late summer. Use 4in/10cm cuttings and remove the soft growing tip. Dip in hormone rooting powder and plant in a cold frame. Overwinter there and move to a nursery bed in autumn the following year.
Seed is stratified during the winter. Place outside in a pot of sand, and sow in early spring in a cold frame. Thin out as necessary. Move to a nursery bed in autumn if seedlings look strong, otherwise wait until the following spring. Move to final positions in autumn.

Rhododendron

LAYERING; SEED

Grows to 15ft/4.5m but many reach no more than 5ft/1.5m. *Rhododendron* hybrids have trusses of white, red, yellow, orange, pink or mauve flowers, depending on the variety. They bloom from early spring to summer, again according to variety. *Azalea*, Japanese hybrids, are evergreen or semi-evergreen shrubs, growing to 5ft/1.5m. They produce white, red, pink, orange or purple flowers in spring.

BERRIED SHRUBS

*B*erberis. Red berries
Chaenomeles speciosa. Yellow fruits
Daphne mezerum. Red or yellow
Ilex aquifolium. Scarlet
Mahonia aquifolium. Blue
Pyracantha coccinea. Bright red or orange red
Rosa. Rugosa roses have large scarlet hips

Pyracantha coccinea, Firethorn produces autumn clusters of bright red or orange berries among its evergreen leaves.

Layering takes place in late spring and is probably the most successful propagation method. Choose a shoot of the current year's growth and make a cut at an angle halfway through the stem, about 12in/30cm from the tip. Wedge open with a matchstick. Peg down the cut section in a ready-prepared hole and cover with soil. Tie the tip to a cane to make it grow upright. Leave until autumn the following year before severing from the parent, but only if there are plenty of roots. Otherwise, wait until the following spring.
Seed is sown in spring, on the surface of a mixture of half peat and half sand. Place in a propagator at 55–60°F/ 13–16°C. Germination takes 5 weeks. Thin out to 1in/2.5cm when seedlings can be handled easily. Gradually harden off and transfer to a cold frame, growing on there until spring the following year. Plant out to a nursery bed and transfer to permanent positions the following year. Flowering occurs about 4 years from sowing.

Ribes

HARDWOOD CUTTINGS

Grows to 8ft/2.4m. *Ribes alpinum*, Alpine Currant, has small, greenish-yellow flowers in spring, followed by red berries.

The variety 'Aureum' has attractive yellow leaves. *Ribes alpinum* rarely grows to more than 5ft/1.5m. *Ribes sanguinem* 'King Edward VII', Flowering Currant, produces clusters of bright crimson five-petalled flowers in spring. It remains a compact 4ft/1.2m. *Ribes speciosum*, Fuchsia-Flowered Gooseberry, has brilliant red flowers in late spring, carried on red stems, followed by red berries.

Hardwood cuttings are taken in autumn. Choose shoots of the current year's growth, about 10in/25cm long. Remove the soft tips and make a slanting cut at the top of the cutting, just above a bud. Make a straight cut at the bottom, just below a bud. Keep all buds on; those below ground will grow to produce a bushier shrub. Plant in a trench in the garden or garden frame, 9in/22.5cm apart, ensuring that the straight cut end of the cuttings is buried. Plant deep enough to leave 2 buds only showing above the soil. Move to permanent positions the following autumn.

Rosa

HARDWOOD CUTTINGS

Rosa species and hybrids, shrubs and climbers, grow between 1–15ft/30cm–4.5m. Commercially, most roses are propagated by budding, but species varieties, old fashioned shrub roses, climbers, and miniature roses can be propagated successfully from cuttings, as can strong growing floribunda. Hybrid tea roses raised from cuttings tend to be weak and short lived.

Hardwood cuttings are taken in early autumn or autumn, from 8in/20cm ripe shoots of the current year's growth, with or without a heel. Remove the soft tip and cut the base just below a bud; if the cutting has a heel, cut away any jagged edge. Cut off all leaves except the top pair. Dip the cutting in hormone rooting powder. Make a V-shaped trench in a sheltered part of the garden and insert the cuttings to most of their depth, about 4in/10cm apart. Fill the trench and firm the soil around them. Leave them until autumn of the following year before moving to their flowering position. Miniature roses will probably be the first to bloom.

Spiraea

HARDWOOD CUTTINGS; SOFTWOOD CUTTINGS

Grows to 4ft/1.2m. *Spiraea × arguta*, Foam of May, produces large clusters of white flowers in late spring. *Spiraea × bumalda* 'Anthony Waterer' is a dwarf shrub, growing to 2ft/60cm, with bright crimson flowers in summer. It may also grow cream to pink variegated leaves.

Hardwood cuttings of various forms except *Spiraea × arguta* are taken in early winter. They should be 10in/25cm long, with a slanting cut at the top just above a bud and a straight cut at the bottom just below a bud. Dip the straight cut in hormone rooting powder and plant, about 6in/15cm deep, in a cold frame or a sheltered part of the garden. Plant out the following autumn.
Softwood cuttings of *Spiraea × arguta* are taken in early summer. Choose side shoots about 4in/10cm long and remove them with a heel. Plant in a pot of half peat and half sand. Place in a cold frame or propagator with gentle heat. Cuttings rooted in a propagator are moved to a garden frame for hardening off. Plant out in autumn of the following year.

Syringa

SEMI-RIPE CUTTINGS; LAYERING; SEED

Grows to 10ft/3m and more. *Syringa vulgaris*, Common Lilac, includes many varieties, notably 'Marechal Foch', with trusses of carmine-rose single flowers, and 'Souvenirs de Louis Spaeth', with trusses of wine-red single flowers. The double variety 'Charles Joly' has purplish red flowers and 'Madame Lemoine' pure white flowers, cream yellow in bud. All flower in late spring and early summer. *Syringa microphylla* is much smaller, growing to 6ft/1.8m. Its rose-pink flowers appear in late spring and early summer and a second flowering may bloom from late summer to autumn.

Semi-ripe cuttings, 4in/10cm long, are taken in late summer. Plant in a mixture of half peat and half sand and place in a propagator at 65°F/18°C. Rooting will take some time. When growing well, harden off and transfer to a garden frame. Plant out to permanent position the following autumn.
Layering takes place in spring and summer if the shrub has produced suitable shoots near the ground. Choose a shoot of the current year's growth. Make an angled cut halfway through the stem, 12in/30cm from the tip. Wedge the cut open with a matchstick. Dig a shallow hole and peg down the cut stem. Cover with soil and support the shoot above ground with a stick, to make it grow upright. If plenty of roots develop by the following spring, sever the layer from the parent plant. If not, leave the layer undisturbed for a time before moving to its permanent place in the garden.
Seed is placed in a pot in early winter and left outdoors to stratify. In spring, snow the seed in a pot of loam seed compost and place in a cold frame. Transplant to a nursery bed in autumn or, if the seedlings are not robust, leave until the following spring. Plant out in the garden in late autumn the following year.

Viburnum

SEMI-RIPE CUTTINGS; LAYERING; SEED

Grows to 6ft/1.8m. In spring, *Viburnum carlesii* produces heavily scented clusters of star-shaped flowers, pink in bud but white when open. The oval, toothed, dark green leaves have downy undersides.

Semi-ripe cuttings are taken in late summer. Remove 4in/10cm stem cuttings and plant in a cold frame. Overwinter there and plant out in autumn the following year.
Layering takes place in spring. Choose a shoot of the previous year's growth and make a cut halfway through the stem, 10in/25cm from the tip. Wedge open with a matchstick. Peg down the cut section in a ready-prepared hole and cover with soil. Support the tip with a cane to make it grow upright. Sever from the parent plant in spring the following year if there are plenty of roots. Otherwise, wait until autumn and plant out to permanent positions.
Seed is sown in spring in a cold frame. The seed will develop roots but will not put out growth above the surface of the compost until spring or later the following year. Thin out when young plants can be handled easily. Plant out in autumn if robust, otherwise wait until spring the following year.

Weigelia

SEMI-RIPE CUTTINGS; HARDWOOD CUTTINGS

Grows to 6ft/1.8m. *Weigelia florida* hybrids produce flowers in varying shades of red and pink, from late spring to summer.

Semi-ripe cuttings are taken from early to late summer. Remove 4in/10cm cuttings and plant in a cold frame. Leave there to overwinter and plant out in autumn the following year.
Hardwood cuttings are taken in late autumn. Remove 10in/25cm cuttings, making a slanting cut just above a bud and a straight cut just below a node. Dip the straight end in hormone rooting powder and plant in a cold frame. Plant out the following autumn.

Shrubs

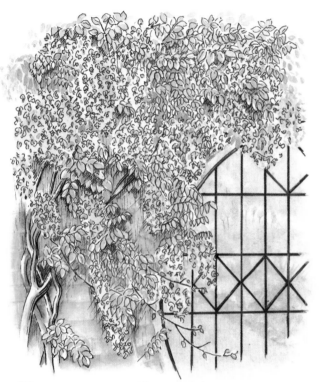

Climbers

A number of climbing plants give the opportunity to experiment with a form of propagation not yet detailed. This is serpentine layering, which is really just a more elaborate form of simple layering.

Serpentine layering allows several new plants to be raised from one shoot. Such shoots must obviously be long, so the method is particularly suitable for climbing and trailing plants. The operation is carried out between spring and late summer. To prepare the ground, mix peat and sand into the soil around the shrub where the layers are to be rooted. This will aid drainage; buried stems may rot in a heavy, wet soil.

Choose young shoots, and, as in simple layering (described on page 56) wound them first to encourage rooting. To do this, make a number of cuts along the stem, at points where the shoot is to be buried. The cut should be about 2in/5cm long; it is made on the slant along the stem and no more than half way into it. Wedge

a matchstick into the cut to keep it open. Any leaves which will be buried should be removed. Then make depressions in the soil ready to bury and peg down the stem near each wound. If preferred, the wounded sections of stem may be layered into pots, using a mixture of equal parts of peat and coarse sand. Layering in pots means less disturbance when transplanting.

If rooting directly into the soil, pin down the layers with bent pieces of galvanized wire and cover with soil. Keep the layers watered in dry weather and free from weeds; hand weed because hoeing might cut the buried stems. Most climbers layered in this way will root within a year. Then cut the rooted parts away from the parent stems and transplant into the nursery bed or straight into the garden.

Pots containing layers can be buried in the ground so that the compost does not dry out quickly in hot weather or freeze too easily in cold. When planting, make sure that the cut in the stem is buried safely below the surface of the compost.

Propagation by **leaf bud cuttings** (see page 110) can be attempted with several climbers, as well as the usual methods of stem cuttings (softwood, semi-ripe and hardwood) and seed. But some ornamental vines offer yet another method to try: **eye cuttings**.

Eye cuttings are small lengths of hardwood stems taken from dormant plants in early winter or winter. The stems may be cut into 1.5in/3.75cm pieces with a single eye — the bud — at the end of the cutting; or they may be cut with the bud in the middle.

For an end eye cutting, make the top cut just above a bud and the lower cut between buds. Dip the end in hormone rooting powder and push the cutting vertically into a pot of loam seed compost. Leave only the bud showing above the surface.

For the alternative type of cutting, take a 1.5in/2.75cm piece of stem with the bud in the middle. Remove a strip of bark the length of the cutting on the other side from the bud. Dust the cut edge with hormone rooting powder and press the cutting on to the surface of the compost. Then bury it under a layer of sand, leaving only the bud showing.

Both types of cutting need placing in a propagator at 75°F/24°C. The warmth will probably cause the eye to grow before rooting has started. Therefore do not transplant before spring, putting them in individual pots of loam potting compost. Harden off gradually before moving to a cold frame, in preparation for planting out. The cuttings grow long stems which at this stage should be supported by a cane.

Serpentine layering

Eye cuttings

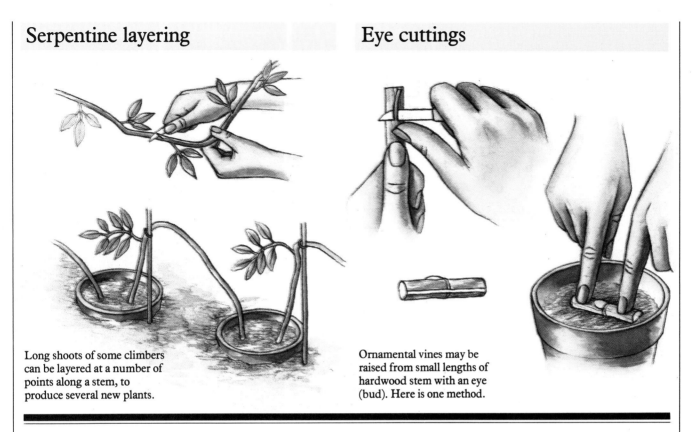

Long shoots of some climbers can be layered at a number of points along a stem, to produce several new plants.

Ornamental vines may be raised from small lengths of hardwood stem with an eye (bud). Here is one method.

Clematis

SERPENTINE LAYERING;
SEMI-RIPE CUTTINGS; SEED

Flowering climbers, mainly deciduous, with twining leaf-stalks which need support. The many species and cultivars grow from 6ft/1.8m to 30ft/9m, giving a succession of flowers from late winter through to early autumn. Among the most popular are *Clematis armandii*, evergreen, with white flowers in spring; *C. macropetala*, pink-flowered in spring; *C. montana*, very vigorous, with white or pink flowers in early summer; *C. orientalis*, the Orange Peel Clematis, yellow in early autumn; *C. tagutica*, bright yellow and *C. flammula*, fragrant yellow in early autumn.

Serpentine layering (see page 64) can be carried out from spring to late summer.
Semi-ripe cuttings are taken in summer or late summer. Clematis cuttings are thought to root more readily if cut half

way between nodes (inter-nodal cuttings). Treat as other semi-ripe cuttings. Root directly into a cold frame, or in pots in the frame, to overwinter.
Seed is sown straight away when ripe, or stratified during winter and sown in early spring or spring. If sown indoors, a little heat only is needed, but seeds will also germinate in a frame or in the open ground in a seed bed, to be transplanted to the nursery bed. Species types of clematis can be grown from seed; cultivars must be propagated from layers or cuttings to come true.

Hedera

SOFTWOOD CUTTINGS; LAYERING

Evergreen self-clinging climber. Can grow to 50ft/15m. *Hedera helix*, Ivy, has 3 to 5-lobed dark green leaves on trailing or climbing stems, but the cultivars have a great range of variegations. *Hedera colchica*, Persian Ivy, has large, heart-

shaped dark green leaves. *Hedera canariensis* 'Variegata' (Gloire de Marengo) with large leathery leaves and cream to white edges, is less hardy.

Softwood cuttings are taken in late spring or early summer. Use stem cuttings 4in/10cm long, making cuts just below nodes. Trim off the bottom 2 leaves and plant several cuttings in a pot of half loam potting compost and half sand. Keep out of direct sunlight. Cuttings should root in 3–4 weeks. When well established, plant where they are to grow.
Layering takes place in late spring or early summer. Ivies often put out aerial roots from growing points (nodes) on the stems. Fill pots with half loam potting compost and half sand. Pin down the tip, about 3in/7.5cm long, of a trailing stem on to the surface of the compost of each pot. When new growth appears from the cutting, cut away from the parent plant and move it where it is to grow. Ivies can also be propagated by serpentine layering (see page 64).

Shrubs

Hydrangea

SEED; SEMI-RIPE CUTTINGS; SERPENTINE LAYERING

Flowering climber, self-clinging by aerial roots; deciduous. Can grow to 30–40ft/9–12m, slowly at first. *Hydrangea petiolaris* has vivid green leaves and in early summer and summer is covered in lace-cap heads of flowers. It is slow to reach the flowering stage.

Seed is sown in early spring or spring, in pots of loam based seed compost. Sow thinly and cover lightly. Place in a propagator at 65°F/18°C. After pricking out the seedlings, start to harden them off before moving to a cold frame for overwintering.
Semi-ripe cuttings are taken in summer and late summer, and planted in a greenhouse or frame. Take cuttings of about 6in/15cm, remove the soft tip, trim the bottom end to just below a node and take off the lower leaves. A temperature of about 65°F/18°C will encourage rooting. Harden off and overwinter in a frame.
Serpentine layering is also possible (see page 64).

Jasminum

SEMI-RIPE CUTTINGS; LAYERING

Flowering climbers which need support. *Jasminum nudiflorum*, Winter Jasmine, grows 12ft/3.6m and more, with bright yellow folowers on bare stems from late autumn to early spring. *Jasminum officinale*, Summer Jasmine, can grow to 30ft/9m. This twining plant needs support for its stems. Trusses of sweetly scented white flowers bloom from summer.

Semi-ripe cuttings are taken in late summer. Use 4in/10cm stem tip cuttings, making the cut just below a node (growing point). Trim off the bottom pair of leaves and insert in a pot of half loam potting compost and half sand. Cuttings will root better if the pot is placed in a plastic bag and kept at around 60°F/16°C, but out of sun. Rooting takes about 4 weeks; the plants can then be hardened off.
Layering often occurs naturally in

Jasminum nudiflorum when one of the long shoots touches the soil. Both simple and serpentine layering (see page 64) are easy to undertake.

Lonicera

SERPENTINE LAYERING; SEMI-RIPE CUTTINGS

Flowering climber. Some varieties are deciduous, some evergreen. Most are vigorous and, with few exceptions, heavily scented. The plants twine and need support for training. *Lonicera periclymenum* 'Belgica', Early Dutch Honeysuckle, has pale rose and yellow blooms from late spring to summer; *L.p.* 'Serotina', Late Dutch Honeysuckle, flowers from summer to early autumn; *L × americana* has white flowers from early summer to early autumn; *L. japonica* 'Halliana' has white blooms from early summer to autumn, and *L.j.* 'Aureoreticulata', yellow flowers from early to late summer.

Serpentine layering takes place between spring and late summer (see page 64).
Semi-ripe cuttings are taken in early summer to summer from riper wood of the current year's growth. Use pieces up to 6in/15cm long. Remove the soft tip, trim the bottom end to just below a node and remove the lower leaves. Dip the ends in hormone rooting powder. Root in a frame or, if possible, start in a propagator between 60–65°F/16–18°C, harden off when roots are growing well, and overwinter in a cold frame.

Passiflora

LEAF BUD CUTTINGS

Flowering climber, with tendrils. Almost evergreen in mild areas but safer grown in an unheated greenhouse in very cold areas. Reaches 10ft/3m or more. *Passiflora caerulea*, Passion Flower, has shiny, dark green, lobed leaves and showy, slightly fragrant, 5-petalled flowers from summer to early autumn.

Leaf bud cuttings are taken in spring. Remove 2 lengths of stem tip, about 10in/25cm. Cut into pieces so that each

Leaf bud cuttings of *Passiflora caerulea*

has a leaf with 1in/2.5cm of stem above the leaf stalk and 2in/5cm of stem below it. Dip the base of the cuttings in hormone rooting powder. Insert in a mixture of half loam potting compost and half sand so that the point where the leaf stalk joins the stem is at compost level. Place in a propagator at 70°F/21°C. New growth should appear after 4–6 weeks, showing that the cutting has rooted. Remove from the propagator and replant in loam potting compost. Harden off gradually before planting outside. Give some protection during winter in the plant's early years. This can be done by loosely piling some straw around the plant.

Parthenocissus

SERPENTINE LAYERING; SOFTWOOD CUTTINGS; SEED

Woody vines, commonly described as Virginia Creepers, notable for the colour of their leaves in autumn. *Parthenocissus henryana* is the most striking, its velvety green leaves turning a brilliant red. *Parthenocissus tricuspidata*, the Boston Ivy, has leaves which turn from light green to red and orange. They grow to 20ft/6m or so, clinging by tendrils and adhesive pads.

Serpentine layering (see page 64) is the easiest way of propagation, because layered vines root so quickly.

Softwood cuttings are taken in early summer and root fairly well in a humid atmosphere to prevent wilting. Cut them 4–6in/10–15cm long and plant in a mixture of loam potting compost and sand.

Hardwood cuttings are taken when the plant is dormant from late autumn. Use pieces 8in/20cm or so long and dip the end in hormone rooting powder. They can be planted directly in the ground, 4–6in/10–15cm apart, and deep enough so that only one or two buds are above ground. However, root disturbance (which they hate) will be lessened if they are rooted in individual pots in a cold frame.

Seed is sown straight away when ripe. If this is not possible, stratify the seeds for 3 months and sow in a seed bed in the spring. Transfer plants to the nursery bed when large enough.

Polygonum

HARDWOOD CUTTINGS; SEMI-RIPE CUTTINGS

Foliage climber, with twining stems. Rampant, easily reaching 40ft/12m, but needs some support at the start. *Polygonum baldschuanicum*, Russian Vine, has large, pale green leaves and from summer is smothered in dense panicles of small flowers, pale cream tinged pink. Invaluable plant for covering eyesores.

Semi-ripe cuttings are taken in summer from woody parts of the current year's growth. Take pieces about 6in/15cm long, preferably with a heel. Remove the soft tip and lower leaves. Dip the heel in hormone rooting powder and insert in pots of seed compost to a depth of about a quarter of their length. Place in a cold frame.

Hardwood cuttings are taken in autumn or late autumn. Use pieces 10in/25cm long. Plant in a cold frame. They will not begin to root until spring, so keep in the frame until the following autumn.

Vitis

EYE CUTTINGS; SERPENTINE LAYERING

Ornamental vines, climbing with tendrils. Vigorous growers, some to 60ft/18m. *Vitis cognetiae*, the most spectacular, has large leaves 10in/25cm long and 6–8in/15–20cm wide. Through summer they are dark green above and downy russet underneath, turning crimson and scarlet in autumn. *Vitis labrusca*, Northern Fox Grape, has sweetly scented flowers, purple fruits, and dark green leaves which turn multi-coloured in autumn. Variety 'Brandt' of the common grape, *Vitis vinifera*, has orange, yellow, scarlet and crimson foliage in autumn. *Vitis vinifera* 'Purpurea' turns from deep red to purple.

Eye cuttings (see page 64) can be taken for all except *Vitis cognetiae*.
Serpentine layering (see page 64) takes place in spring. All can be layered.

Wisteria

LAYERING (SIMPLE OR SERPENTINE)

Flowering climbers, with twining stems which need support. Growth, according to species, is between 25ft/7.5m and 60ft/18m. *Wisteria sinensis*, Chinese Wisteria, twines anti-clockwise. Its long racemes of heavily scented deep mauve flowers bloom in late spring and early summer. *Wisteria floribunda*, the less vigorous Japanese Wisteria, twines clockwise. Violet racemes (or white in the variety 'Alba') flower a few weeks after *Wisteria sinensis*.

Simple or serpentine layering (see pages 55 and 64) are the simplest ways of propagating wisteria. Growing from seed is liable to produce unsatisfactory plants. It will take time — even up to 20 years — from sowing to flowering. Even vigorous plants raised vegetatively are not quick to flower.

WHAT'S IN A NAME?

The Latinized name of a plant identifies that plant throughout the world; it is a kind of identity tag.

The first name shows to which **genus** or family the plant belongs. A genus includes all plants which have the same **botanical** characteristics, even though they may not look alike. (For example, *Passiflora* is the name of the genus.)

The second name shows to which **species** within the genus the plant belongs. All plants of the same species will – from their appearance – be recognized as belonging to that species. (For example, *Passiflora caerulea*, *Passiflora edulis*, and so on.)

The first two names of a plant are usually printed in italics.

There may be some variations within the species, however — slightly different leaves or flowers, for instance. When there are such differences the plant is given a third name, and this identifies the variety. These variations do occur naturally and this fact is usually shown by printing the name in italics.

Variations are more often bred into a

Passiflora caerulea

plant, however. If so, the plant is described as a cultivated variety or 'cultivar'. The fact that it is a cultivated variety is usually shown by printing the name in roman and not italic type. (For example, *Passiflora caerulea* 'Constance Elliott', which is pure white.)

Bulbs and Corms

There are more than 10,000 species of the main families of bulbs, corms and tubers, and not just snowdrops, crocus, daffodils, hyacinths, tulips and gladioli, as many gardeners seem to believe. They grow in all shapes and sizes, all shades and scents, and they fit in any part of the garden. Planted on their own, in formal beds, they make a great show in spring and summer. They also look at home grown in grass, among shrubs and trees, and in herbaceous borders. Miniature varieties are an essential part of alpine gardens. With careful planning they could be planted as a bulb border to produce flowers virtually all year round; unfortunately they are seldom used in this way. Similarly, they might be used to create a scented garden — about 200 potential candidates exist.

This range of splendour is seen above ground, but what distinguishes the plants in this section from other plants is the way they grow underground. Most are bulbs and corms, with a few tubers and rhizomes. They have comparable root systems, which are unlike those of other plants, and it is their below-ground growth which provides most of the material for propagation and offers interesting methods to be tried out.

There are also a few and rather odd ways of propagating these plants above ground. And then there are seeds. There is no great difficulty in raising bulbs from seed except for the problem of time, testing the gardener's patience.

Who but the most besotted of devotees could contemplate sowing the seeds of *Amaryllis belladonna*, knowing that eight years or more must elapse before there is any hope of seeing a flower? But not all bulbs are

For best results when planting bulbs, corms and tubers, always choose plump, healthy-looking individuals. Take advantage not only of the selection of familiar spring bulbs available, such as these daffodils, tulips and crocus, but also of summer and autumn flowering varieties.

Bulbs and Corms

so dilatory in reaching flowering size, even from seed, and other forms of propagation are faster in producing results. By comparison, propagation from rhizomes and tubers seems almost speedy.

Raising bulbs, corms and tubers from seed is explained on page 79, where there is a table of those which can be expected to flower within two to four years of sowing. The other main forms of propagation for these groups of plants are given below.

BULB OFFSPRING

Simple division. Bulbs propagate themselves in this way and the gardener can leave them to do so for some time before intervening. Some bulbs, daffodils for instance, produce an offset at the base each year. This grows alongside the parent for a year or two, then separates itself from the parent and in time will begin to flower. Other bulbs, tulips for example, disintegrate after flowering, leaving behind a new bulb and some small ones. Over a period, therefore, clumps of bulbs can become overcrowded, with a resulting loss of vigour. To avoid this, the bulbs can be lifted every three or four years, separated, and planted in smaller

groups, preferably not in the same spot. Waiting a few years in between each operation gives the offspring a chance to grow undisturbed to flowering size. Frequent disturbance delays it.

Removal of bulblets and cormlets. These are miniature bulbs or corms growing at the base of the parent. The original bulbs or corms are usually lifted when their leaves have died down. Their offspring are then removed, and planted. Expect to wait two or three years before the young bulblets and cormlets produce their own flowers.

Removal of bulbils. These are very small bulbs found on the stems of some lilies, usually at the leaf axils. They are carefully removed in early autumn, planted in trays of loam potting compost, kept growing in a cold frame for two years and then planted out in the autumn to their flowering position. They should bloom the following year — a wait of three years.

Planting bulb scales is a method often used with lilies. A few of the outer leaf scales are broken off, or cut as close as possible to the base of the bulb. The scales are then planted upright in compost and kept in the warmth. After a few months, bulblets begin to grow from the base of the scales; they develop roots

Simple division

Daffodils are among the bulbs which grow an offset each year. If left, the offsets will form clumps.

To avoid overcrowding, lift the clumps every 3 or 4 years and replant the young bulbs in smaller groups.

Removing bulblets and cormlets

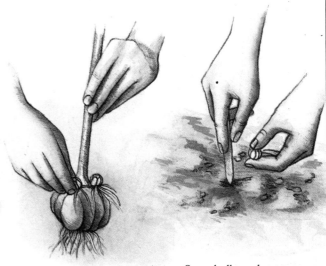

Some bulbs and corms produce miniatures of themselves. If removed and replanted they will flower in 2–3 years.

and then leaves. In autumn, after the leaves have died down, the plants are taken from the pot. The bulblets are removed from the scales and replanted in individual pots. Flowering takes another two years.

Scoring the base of a bulb is an operation often used to encourage hyacinths, which are shy to propagate, but it can also be done with scillas and some lilies. Two incisions in the form of a cross are made in the base of the plant, and from these bulblets will appear. Hyacinths can be grown to flowering size by this method in three to five years. For full details, see the entry hyacinth, page 75.

RHIZOMES AND TUBERS

Division of rhizomes is used for some rhizomatous iris, for example. When the rhizomes begin to sprout in spring, they are lifted and cut into pieces, each one with a leaf and some fibrous roots.

Division of tubers, as used for anemones and cyclamen, involves lifting tubers after they begin to sprout in the spring. These are cut into several pieces, each one with an 'eye' from which the new plant will grow.

Bulb scales

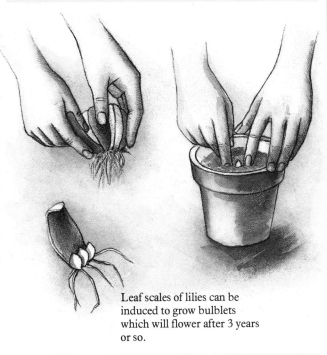

Leaf scales of lilies can be induced to grow bulblets which will flower after 3 years or so.

Removing bulbils

Bulbils grow on the stems of some lilies. Remove carefully and plant, but it will be some 3 years before they flower.

Dividing rhizomes

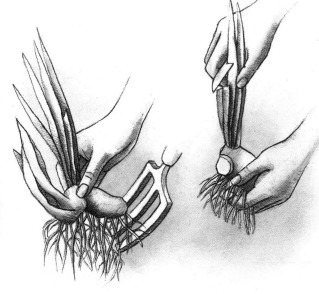

The rhizomes of some iris are divided in spring after they have sprouted. Trim before planting.

Bulbs and Corms

Allium

OFFSETS (BULBLETS)

Allium affatunense

Grows to 3ft/90cm. Plant bulb 3in/7.5cm deep. *Allium affatunense* has wide, strap-shaped leaves. In early summer, large, globe-shaped heads of rose-purple, star-shaped flowers bloom on 3ft/90cm stems. *Allium karataviense* is chiefly grown for its broad velvety blue-green leaves. Globular heads of pinkish white flowers bloom on a 9in/22.5cm flower spike. *Allium oreophilum* (syn. *Allium ostrowskianum*) grows 6–9in/15–22.5cm, producing heads of pink star-shaped flowers in the variety 'Zwanenburg'.

Offsets can be detached when plants are lifted in early autumn. These small bulblets grow from the base of the bulb. Plant them to a depth of 1in/2.5cm in pots or boxes of half loam potting compost and half sand. Allow similar spacing between bulblets. Place in a cold frame. Bulblets should reach flowering size in 2 years and can then be planted in the ground at a depth of 3in/7.5cm.

Alstroemaria

DIVISION OF TUBERS

Grows to 3ft/90cm. Plant tuber 6in/15cm deep. *Alstroemaria aurantiaca*, Peruvian Lily, has short, narrow, pointed leaves

and funnel-shaped, orange-yellow flowers in summer months. The variety 'Moerheim Orange' has larger and deeper coloured flowers. *Alstroemaria* 'Ligtu' hybrids produce 2ft/60cm stems of pink, orange or yellow flowers during summer.

Division of tubers takes place in spring. The roots are brittle and great care must be taken. Tubers and roots grow deep in the ground, so a fork will be required to separate and lift them. Plant divided tubers to a depth of 6in/15cm, about 10in/25cm apart.

Amaryllis

OFFSETS

Grows to 2ft/60cm. Plant bulb 6in/15cm deep. *Amaryllis belladona*, Belladona Lily, has loose clusters of trumpet-shaped, pale pink flowers on long stems in autumn. Strap-shaped leaves follow the flowers.

Offsets are removed in late spring when repotting, which takes place every 4–5 years. It is then you will discover whether offsets have developed. If they have, plant them in a pot of equal parts of loam potting compost, leaf mould and sand, with the tip of the bulb just showing above the surface. Keep the pots in a frame or greenhouse while growing on. Bulbs planted in the ground are best left undisturbed if flowering is not to be affected.

Anemone

SEED; DIVISION OF CORMS OR TUBERS

Grows to 3ft/90cm. Plant corms and tubers 2in/5cm deep. *Anemone coronaria*, Poppy Anemone, has deeply cut, parsley-like foliage and blue, purple, white or red flowers in spring or early summer. The variety 'St Brigid' has large double and semi-double flowers in mixed colours and 'The Bride', white single flowers. It grows to 12in/30cm. *Anemone japonica*, Japanese Anemone, has large pink, mauve and white single and semi-double flowers in late summer, growing to 3ft/90cm. *Anemone narcissiflora*, reaching 18in/45cm, produces clusters of white to pink flowers from late spring to early

summer. *Anemone nemorosa*, Wood Anemone, grows to 8in/20cm, with white to pink flowers in spring.

Anemone coronaria

Seed is sown in spring. Prepare a tray with loam seed compost and scatter the seed evenly over the surface. Do not cover with compost, but fit a sheet of glass over the tray and keep at a temperature between 55–60°F/13–16°C. When seedlings can be handled easily, transplant to outdoor beds.

Division takes place in early autumn after foliage has died down. Corms and tubers can be lifted and cut into several pieces, each with one or more eyes, which are the growing points. Dust the cut pieces with fungicide powder and store in dry sand or peat until planting time, with the exception of spring flowering corms and tubers. Once these have been dusted with fungicide, leave in the warmth to scar over for 48 hours and then plant in open ground. Plantings of corms and tubers can be made in autumn, spring and summer for flowering succession.

Brodiaea

OFFSETS

Grows to 2ft/60cm. Plant bulb 6in/15cm deep. *Brodiaea californica* grows to 18in/45cm, producing loose clusters of mauve-rose flowers in summer. *Brodiaea coccinea*, Californian Firecracker, produces in summer long-tubed, crimson flowers tipped green and yellow. *Brodiaea laxa* has clusters of violet-blue, tubular flowers, opening to a star shape in summer.

Brodiaea laxa

Offsets are removed when flowers have died down in autumn. Bulbs can be lifted and any offsets detached. Plant 1in/2.5cm deep in pots or boxes of half loam potting compost and half sand, with similar spacing between bulblets. Place in a cold frame. After 2 years, plant out to a depth of 3in/7.5cm.

Camassia

OFFSETS

Grows to 3ft/90cm. Plant bulbs 3in/7.5cm deep. *Camassia esculenta*, Common Camass, has narrow, strap-shaped leaves about 12in/30cm long and 3ft/90cm stems bearing clusters of blue to mauve star-shaped flowers in early summer. *Camassia leichtlinii* has tall flower spikes with loose clusters of creamy-white, star-shaped flowers in summer.

Offsets are removed when foliage has died down in autumn. Bulbs can be lifted to see if any offsets have appeared round the base of the bulb. If so, detach and plant in pots or boxes of half loam potting compost and half sand to a depth of 1in/2.5cm. Place in a cold frame and plant out in the open a year later at a depth of 3in/7.5cm.

Chionodoxa

OFFSETS

Grows to 6in/15cm. Plant bulbs 3in/7.5cm deep. *Chionodoxa luciliae*, Glory of the Snow, produces in spring spikes of pale blue, star-shaped flowers with white centres. There are also varieties with pink flowers. *Chionodoxa sardensis* is smaller, with clusters of deep blue flowers in early spring.

Offsets are removed as soon as the flowers have died down. It is best to divide groups of bulbs into several large pieces, each with several bulbs, rather than detach single offsets. Individual bulbs will take several years to flower, whereas clumps of bulbs should produce some flowers from the following year onwards. Replant the divided pieces at once.

Colchicum

DIVISION OF CORMS

Grows to 12in/30cm. Plant corms 4in/10cm deep. *Colchicum autumnale*, Meadow Saffron, has crocus-shaped, mauve-pink flowers, up to 6in/15cm in height, in early autumn. Lance-shaped leaves which follow the flowers die down in the following late spring. Some forms have white flowers. *Colchicum luteum* has yellow flowers, up to 4in/10cm tall, of similar shape, which appear in early spring. *Colchicum speciosum* grows to 12in/30cm, producing pale lilac to deep purple flowers in early autumn. There is also a white flowering form.

Division of corms takes place every

Colchicum autumnale

third year in summer. Lift clumps, and from that which has been the latest to flower (and which should now be dying back) break away new corms. Flowering-size corms are usually about 8in/20cm across. Smaller ones will take a year or two before they flower. Replant immediately in the ground.

Convallaria

DIVISION OF RHIZOMES

Grows to 8in/20cm. Plant rhizomes 3in/7.5cm deep. *Convallaria majalis*, Lily of the Valley, has lance-shaped, dark green leaves and drooping heads of small, fragrant, white bells in late spring.

Division of rhizome, usually referred to as the crown, takes place in autumn. Lift and divide the crown, but not into many small pieces as these will take some time to produce a good show of flowers. Two or three large pieces will ensure flowering the following spring. Plant 6in/15cm apart.

Bulbs and Corms

Crinum

OFFSETS

Grows to 3ft/90cm. Plant bulb with neck showing at compost level. *Crinum × powellii* is a cross between *Crinum bulbispermum* and *Crinum moorei*, producing in summer light green, strap-shaped leaves and slightly drooping heads of white to pink lily-like flowers. Will grow outdoors only in warm, basically frost-free areas: otherwise grow in pots for summer show and overwinter under glass.

Offsets are removed in spring when repotting. Remove plant from its pot and break away the compost. Older bulbs should have produced several offsets round the base. These can be detached and planted in a mixture of loam, leaf mould and sand, with the neck showing above compost level. Flowering will begin in about 3 years.

Crocosmia

OFFSETS

Grows to 2ft/60cm. Plant corm 5in/12.5cm deep. *Crocosmia × crocosmiiflora*, Montbretia, bears stems of star-shaped, yellow to deep orange flowers in late summer.

Offsets are removed in spring. Lift clumps of corms every 3 to 4 years to be divided. It is best to divide clumps into several large pieces rather than break off individual cormlets which would take about 2 years to come into flower Replant divided pieces immediately.

Crocus

OFFSETS

Grows to 6in/15cm. Plant corms 3in/7.5cm deep. *Crocus chrysanthus* has been developed in many varieties. Cup-shaped, white, yellow, blue, deep purple, often bicoloured, streaked flowers are produced in early spring. Narrow, grass-like leaves have a central white streak along their length. *Crocus speciosus*, Autumn Crocus, has pale purple, cup-

CYCLAMEN

Hardy cyclamen, such as *C. europaeum* and *C. neapolitanum* (see below), are grown outdoors, often under trees or among shrubs. Undisturbed, they will often seed themselves, and colonise large areas. The cyclamen grown as a houseplant is *C. persicum* (see page 116). Though not hardy outdoors, it is more likely to die from heat indoors. It will survive better at temperatures in the range 45-55°F/7-13°C.

shaped flowers in autmn. White flowering varieties are also available.

Offsets are removed every 4 to 5 years when the foliage has died down after flowering. Break away single new corms from the old decaying ones and replant immediately at a depth of 3in/7.5cm. Spring-flowering corms can be stored in sand and planted out again in the autumn.

Cyclamen

DIVISION OF TUBERS

Grows to 6in/15cm. Plant tubers 1in/2.5cm deep. *Cyclamen europaeum*, Sowbread, has round leaves with silvery markings and shuttlecock-shaped, deep pink flowers during summer months. *Cyclamen neapolitanum* produces mauve-pink flowers during summer months. shaped leaves have silver markings.

Division takes place in late spring, but preparation begins when flowering is over. Lift the tubers, dust with a fungicide powder, wrap in newspaper and store in a frost-free place until late spring. Cut the tubers into 2 or 3 pieces, each with a growing eye. Dust with fungicide powder and leave in the warmth for 48 hours to scale over. Plant each piece in a pot of loam potting compost and, when a new shoot has appeared, plant out to flowering position.

Eranthis

DIVISION OF TUBERS

Grows to 3in/7.5cm. Plant tuber 2in/5cm deep. *Eranthis hyemalis*, Winter Aconite, produces in spring cup-shaped, golden-yellow flowers, similar to buttercups. It has bright green, deeply incised leaves.

Division takes place in autumn. Divide only old tubers which have been in the ground for at least 5 years. Younger, smaller tubers will produce poor flowering plants for several years until they become well established again. Lift one or two older tubers each year in autumn and divide into several pieces, each with growing points. Dust the pieces with fungicide powder and leave in the warmth for 48 hours for the cut areas to scar over. Plant out in flowering positions to a depth of 2in/5cm.

Erythronium

DIVISION

Grows to 10in/25cm. Plant tubers 4in/10cm deep. *Erythronium revolutum* 'Pagoda', Trout Lily, has dark green, lance-shaped leaves, mottled brown. Pale yellow flowers with flared back petals and a brown central ring appear in spring. The variety White Beauty has white flowers with a yellow centre.

Tubers can be lifted in late spring when flowering is over, although clumps of tubers are probably best left undisturbed. If dividing, split the clumps into 2 or 3 large pieces and replant immediately, at a depth of 4in/10cm.

Freesia

SEED; OFFSETS

Grows to 18in/45cm. Plant corms 1in/2.5cm deep. *Freesia refracta* was the original for the many varieties now developed, with white, yellow, orange, pink, or blue fragrant, funnel-shaped flowers produced from winter to spring.

Seed is sown in spring. This is the easiest method of propagation; from sowing to flowering takes 9 months. Soak the seed in warm water for 24 hours. Fill a 6in/15cm pot with a mixture of loam, peat and sand and sow 6–8 seeds. Lightly cover with compost and place in a propagator at 65°F/18°C. Germination takes about 5 weeks. Do not transplant seedlings. Place outdoors during summer and then bring into the greenhouse in early autumn before the first frosts.

Offsets are removed in spring. This method is best for increasing a stock of a particular strain or colour. Lift the corms and detach individual small corms which may have developed. Plant several to a pot, using the mixture recommended above, and keep under glass until summer. Corms should flower by the end of summer.

Fritillaria

OFFSETS

Grows to 3ft/90cm. Plant *Fritillaria imperialis* bulb 6in/15cm deep and all others 3in/7.5cm deep. *Fritillaria imperialis*, Crown Imperial, grows to 3ft/90cm, producing a cluster of bell-shaped, yellow, red or orange-red flowers in spring, arranged in a circle at the top of the flower stalks and topped with a rosette of leaves. *Fritillaria meleagris*, Snake's Head, growing to 18in/45cm, produces in spring large, drooping, bell-shaped white and purple flowers in varying shades with chequered markings carried on single stems. *Fritillaria pallidiflora* has fleshy stems with strap-shaped leaves and yellow drooping bells in spring.

Offsets are removed in autumn, when bulbs may be lifted. Plant immediately in the ground where they are to flower, which will be about 18 months later.

Galanthus

OFFSETS

Grows to 6in/15cm. Plant bulbs 4in/10cm deep. *Galanthus nivalis*, Snowdrop, has sword-shaped, narrow leaves. White pendant flowers, with central petals tipped green, bloom from late winter to early spring.

Offsets are removed in late spring, before the foliage has finally died down. But leave bulbs in the ground for about 5 years before attempting to divide them. Lift the clumps and divide into groups of 4 bulbs. Plant immediately at a depth of 6in/15cm.

Gladiolus

CORMLETS

Grows to 3ft/90cm. Plant corms 4in/10cm deep. *Gladiolus* hybrids, Sword Lilies, have tall spikes of closely packed, trumpet-shaped, red, orange, yellow, white, pink, blue or purple flowers from spring to early summer.

Cormlets are removed in autumn when the leaves begin to turn yellow and die down. Lift the corms with foliage still attached. Tie the leaves together in small groups and hang in an airy place to dry out. Cut away the foliage and remove any cormlets that have formed. Store in a frost-free, airy place at about 45°F/7°C. The following spring, soak the corms in tepid water for 24 hours and then plant outdoors to a depth of 2in/5cm and about 3in/7.5cm apart. Lift again in autumn and store over winter. Flowering size is reached after 2 years.

Haemanthus

OFFSETS

Flower spike and leaves reach up to 2ft/60cm. Plant bulb 4in/10cm deep. *Haemanthus coccineus*, Blood Lily, has a stout, 8in/20cm stem bearing red bracts and surrounding the closely-packed, red flowers in late summer. Broad, strap-shaped leaves follow the flowers. *Haemanthus katherinae* has tall stems bearing globular clusters of star-shaped, red flowers in summer. Wavy-edged oblong leaves are also produced.

Offsets are removed in spring. Both species mentioned above are greenhouse plants grown in pots, but they can be placed outside in summer. When flowering is over and leaves have died down, store the bulbs dry in their pots in a frost-free place. In spring, when the plant starts into growth, remove the soil ball from the pot, break away the compost and remove any offsets which have formed. Plant offsets in a 3in/7.5cm pot, using a mixture of loam, peat and sand, and treat as a mature bulb.

Hyacinthus

OFFSETS

Grows to 10in/25cm. Plant bulb 6in/15cm deep. *Hyacinthus orientalis* is the original for the many varieties now available. In spring, a flower spike appears with clusters of waxy red, pink, purple, blue, white or yellow flowers ranged along most of its length. The leaves are strap-shaped.

Offsets are slow to appear and require helping along the way. In autumn, score the bases of mature bulbs with 2 lines, about 0.5in/1.25cm deep, at right angles to each other. Leave the bulb in a warm, dry place for 3 days to allow the cuts to open up and scar over. Dust the cut area with a fungicide surface of a pot of slightly damp sand, cut area uppermost. Keep warm and dry — an airing cupboard is ideal. Make sure the sand does not dry out completely. After about 12 weeks, small bulblets should appear from the cut surface. Plant the bulb in a pot of loam potting compost with added sand, still with the cut area facing upwards. Cover with the mixture so that the bulblets are just below the surface. Harden off and place in a greenhouse or frame. In spring, shoots will appear. Allow them to grow on until early summer. Then remove the bulb from the pot and separate the individual bulblets. Plant in a bed until flowering size is reached. This can take from 3 to 5 years.

Bulbs and Corms

Ipheion

DIVISION (OFFSETS)

Grows to 6in/15cm. Plant bulbs 3in/7.5cm deep. *Ipheion uniflorum* has narrow, grass-like leaves and, in spring, large, star-shaped deep blue flowers, several to a stem.

Offsets are removed in autumn when the bulbs are lifted. Plant in a box of half loam potting compost and half sand and grow in a cool greenhouse or cold frame for 2 years. Then plant out in autumn to a sunny, protected spot, planting the bulbs 3in/7.5cm deep. Flowering size will be reached in a further 2 years. In cold areas the plants can remain in a greenhouse, several bulbs to a pot.

Iris

OFFSETS; DIVISION OF RHIZOMES

Grows to 3ft/90cm. Plant bulbs 4in/10cm deep and rhizomes 1in/2.5cm deep. *Iris reticulata* (bulb) has narrow, grass-like leaves and violet or yellow flowers with flaring petals in spring. It grows to 6in/15cm. *Iris xiphioides* (bulb) was the original for the many varieties of English irises bearing flowers of varying shades of purple and pure white in summer. *Iris germanica* (rhizomes), Bearded or German Irises, have been used to develop many varieties in almost every imaginable

colour. The bearded flowers appear in early summer.

Offsets are removed when the leaves begin to die down. Lift groups of bulbs and leave them to dry out in the sun. Detach offsets and plant them in flowering positions in autumn.

Iris reticulata

Division of rhizomes takes place in summer, when flowering is over, every 4 years. Lift the plant and cut the rhizome into pieces about 3in/7.5cm long, each with a group of leaves and plenty of roots. If planted as they are, the pieces will be top heavy, so cut the leaves down to 5in/12.5cm. Plant each piece immediately in its flowering position, with the top surface of the rhizome showing just above the soil.

Ixia

DIVISION (OFFSETS)

Grows to 12in/30cm. Plant corms 3in/7.5cm deep. *Ixia* is the African Corn Lily, and there are many varieties. The plant has narrow, lance-shaped leaves and flowers which appear in early summer. Spikes of closely packed, six-petalled flowers bloom in white and shades of yellow, cream, red and pink. Ixias are suitable for growing outdoors only in mild areas. Elsewhere grow in pots in a slightly heated greenhouse.

Offsets are removed after flowering is over. Lift the corms in autumn and remove any cormlets which have formed around the base. Plant in boxes of half loam potting compost and half sand and grow for 2 years in a frame or cool greenhouse. In mild areas, plant in a sunny spot outdoors, to a depth of 3in/7.5cm. In cold areas, plant up to 8 corms to a pot of half loam potting compost and half sand and keep in a slightly heated greenhouse. Keep the compost dry in late summer and early autumn as the foliage dies down.

Ixiolirion

DIVISION (OFFSETS)

Grows to 12in/30cm. Plant bulbs 3in/7.5cm deep. *Ixiolirion pallasii* has narrow, grass-like leaves and loose clusters of purple, trumpet shaped flowers in early summer. Needs a warm, sunny position and some winter protection in colder areas.

Offsets are removed when flowering is over. Lift the bulbs and remove any offsets which have developed. Plant singly in 3in/7.5cm pots of half loam potting compost and half sand and grow in a cool greenhouse or frame for 2 years. Then plant out 3in/7cm deep in early autumn to a sunny flowering spot. In cold areas, cover with cloches during the winter. Flowering size should be reached after a further 2 years.

Lachenalia

OFFSETS.

Grows to 12in/30cm. In pots, plant the bulb just below the surface of the compost. *Lachenalia tricolor*, Cape Cowslip, has spikes of tubular flowers, yellow, red and green, in spring, and broad, strap-shaped leaves. Best grown in a cool greenhouse. Avoid frosts.

Offsets are removed in early autumn before bulbs are started into growth. Remove the soil ball from the pot and break away the compost. Remove any offsets and plant in a 3in/7.5cm pot of half loam potting compost and half sand.

Grow on for 2 years and then plant several bulbs to a large pot, using a mixture of two parts loam potting compost, one part leaf mould and added sand for drainage.

Lilium

OFFSETS; BULBILS; SCALES

Grows to 5ft/1.5m. Plant bulbs 6in/15cm deep, but *Lilium candidum* 1in/2.5cm deep. *Lilium candidum*, Madonna Lily, produces white, trumpet-shaped, fragrant flowers with golden stamens in early summer. *Lilium regale,* Regal Lily, has white, trumpet-shaped flowers with yellow throats tinged maroon on the outer surface in summer. *Lilium tigrinum*, Tiger Lily, has deep orange, spotted flowers with curved back petals in summer.

Offsets are removed in early autumn. Lift clumps of bulbs and carefully divide them, taking care not to damage the scales. Replant individual bulbs in flowering positions immediately.
Bulbils are removed in early autumn. Several species, including *Lilium tigrinum*, produce small bulbils above ground at the leaf axils, the point where the leaves join the stem. When they have ripened in early autumn, remove from the stem and plant in pots or deep trays of loam potting compost, 1in/2.5cm deep and with similar spacing between bulbs. Place in a cold frame and leave for 2 years. Remove any flowers that form. Plant in autumn.
Scales are removed in early autumn. Lift bulbs and carefully break away the larger outer scales. Plant them upright in loam potting compost, with the tip of the scale just above the compost surface. Place in a propagator at 65°F/18°C. In 3 to 4 months, bulblets should develop at the base of the scale. Harden off plants when leaves begin to show. After leaves have died down, remove the soil ball from the pot and gently separate bulblets from the scale. Treat in the same way as bulbils.

Muscari

OFFSETS

Grows to 18in/45cm. Plant bulb 3in/7.5cm deep. *Muscari armeniacum*, Grape Hyacinth, has thick heads of bright blue, bell-shaped flowers in spring, up to 8in/20cm tall. It produces grass-like leaves. *Muscari comosum*, Tassel Hyacinth, has narrow, strap-shaped leaves and flower stems up to 18in/45cm long, bearing blue flowers with a plume-like effect.

Offsets are removed in early autumn after leaving bulbs undisturbed for 4 to 5 years. Lift and divide clumps into pieces, each with several bulbs. Replant immediately, 3in/7.5cm deep.

Narcissus

OFFSETS

Grows to 18in/45cm. Plant bulb 4–6in/10–15cm deep. *Narcissus bulbocodium,* Hoop Petticoat, has rush-like, narrow leaves with yellow flowers shaped like a hoop petticoat, about 6in/15cm tall, in spring. *Narcissus* hybrids include trumpet daffodils, large and small cupped narcissi and double narcissi—all producing plain white and yellow trumpet-shaped flowers in spring, or combinations of white and yellow and yellow and orange.

Offsets are removed in late spring or early summer, every 4 to 5 years, when leaves have died down. Lift clumps of bulbs and divide into groups of 3 or 4 bulbs, replanting immediately about 6in/15cm apart.

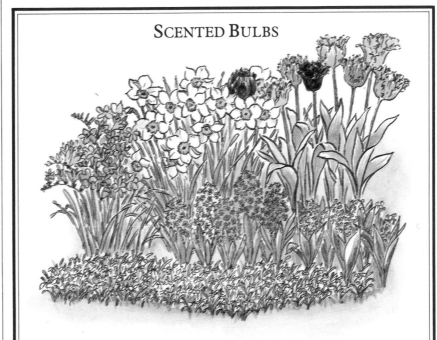

SCENTED BULBS

There are too many scented bulbs to be listed here. As a very general rule, bulbs which have white, pale yellow and pink flowers are most likely to be strongly scented (as is the case with other plants). These are the predominant colours of bulbs flowering in autumn, winter and early spring months.

Scarlet and blue flowers are usually the least heavily scented, although hyacinths and some iris are obvious exceptions. There is an enormous range of fragrance in flowers, depending on their chemical substances. Flowers exist which smell of cucumber, lemon, ripe plums, honey, foxes, dead rats and rotting meat. Most of these 'perfumes' are represented in bulbs.

Bulbs and Corms

Ornithogalum

OFFSETS

Grows to 18in/45cm. Plant bulb 3in/7.5cm deep. *Ornithogalum thyrsoides*, Chincherinchee, has dense clusters of cup-shaped, opening to star-shaped, white flowers in summer, carried on 18in/45cm stems. *Ornithogalum umbellatum*, Star of Bethlehem, has star-like flowers on 6in/15cm stems in spring.

Offsets are removed in autumn. Lift bulbs and break away individual offsets from the base of the bulb. Plant in deep trays or boxes of two thirds loam potting compost and one third sand. Grow on for 2 years and then plant out to flowering positions.

Oxalis

DIVISION

Grows to 4in/10cm. Plant tubers 3in/7.5cm deep. *Oxalis adenophylla* has bright green leaves, divided into several segments radiating from the stalk in a fan shape. Bowl shaped, five petalled pink flowers with a darker eye appear from late spring to early summer. *Oxalis deppei*, Four-leaf Clover, has clover shaped leaves and rose red, flat flowers from early to late summer. *Oxalis brasiliensis* has clusters of crimson red flowers in late spring and early summer.

Division of tubers takes place in spring. Lift and divide into sections, replanting immediately 3in/7.5cm deep. These plants do best in a limey soil.

Puschkinia

DIVISION (OFFSETS)

Grows to 6in/15cm. Plant bulbs 3in/7.5cm deep. *Puschkinia libanotica*, Striped Squill, has narrow, strap shaped leaves and spikes of pale blue flowers in early spring. *Puschkinia libonotica alba* bears white flowers.

Offsets are removed in autumn when the bulbs are lifted. Plant in boxes of half loam potting compost and half sand.

Grow in a cool greenhouse or frame for 2 years. Plant out in autumn. Flowering size is reached in 18 months. Bulbs can also be planted in pots, buried half to their rims outdoors. In winter bring into a greenhouse to flower.

Ranunculus

OFFSETS

Grows to 18in/45cm. Plant tuber 2in/5cm deep. *Ranunculus asiaticus*, Turban Ranunculus, has double and semi-double flowers in early summer, in red, pink, orange, salmon or yellow. In mild, frost-free areas the plants can remain in the ground, but elsewhere they should be lifted in autumn and stored over winter.
Offsets are removed in autumn when foliage dies down. Lift tubers and examine them to see if offsets have grown. If so, plant in boxes of two thirds loam potting compost and one third sand, with the claw-like ends pointing downwards. Place in a greenhouse. Plant out the following autumn in frost-free areas, or 18 months later in spring should plants require winter protection.

Scilla

OFFSETS

Grows to 8in/20cm. Plant bulb 3in/7.5cm deep. *Scilla sibirica*, Squill, has strap-shaped leaves and blue, bell-shaped, drooping flowers in early summer.

Offsets are removed in summer every 4 years when foliage has died down. Lift tubers and divide clumps into pieces with 4 or 5 bulbs. Replant immediately.

Sparaxis

DIVISION (OFFSETS)

Grows to 10in/25cm. Plant corms 3in/7.5cm deep. *Sparaxis tricolor*, Harlequin Flower, has narrow, lance-shaped leaves. From late spring to early summer it carries spikes of red to orange, six-petalled flowers with yellow centres. Other varieties have white, pink, scarlet and yellow flowers.

Offsets are removed in early autumn. Lift corms and remove any offsets which have grown round the base. Plant in boxes of half loam potting compost and half sand. Keep in a cool greenhouse for 2 years until flowering size is reached. In warm areas, move outdoors to flower, planting out in early autumn to a depth of 3in/7.5cm and 3in/7.5cm apart. In cold areas, plant several corms in pots of half loam potting compost and half sand and place in a slightly heated greenhouse to flower.

Sternbergia

OFFSETS

Grows to 5in/12.5cm. Plant bulb 5in/12.5cm deep. *Sternbergia lutea*, Lily of the Field, has narrow, strap-shaped leaves and deep yellow, cup-shaped flowers, similar to crocus, in autumn.

Offsets are removed after flowering when bulbs have been left undisturbed for about 5 years. Lift when leaves have died down and remove any offsets. Plant them individually in 3in/7.5cm pots of two thirds loam potting compost and one third sand. Place in an unheated greenhouse for 2 years. Plant out to flowering positions in early summer. Flowering size is reached after about 4 years.

Tulipa

OFFSETS

Grows to 18in/45cm. Plant bulb 6in/15cm deep. *Tulipa forsteriana* has been the original from which many hybrids have been developed. Bright red, orange, orange-scarlet, yellow, or white, cup-shaped flowers bloom in spring.

Offsets are removed after flowering. Lift the bulbs and detach any offsets, leaving them to dry out. Plant in autumn in deep trays, or a bed about 3in/7.5cm deep. Trays should be left outside. Feed regularly in spring and summer and remove any flower heads to concentrate energy on producing a large bulb. Lift and dry out the bulbs in summer. Repeat the process for a further year, after which they can be planted out to final flowering positions.

BULBS FROM SEED

Raising bulbs from seed is the cheapest way to grow them, but also the slowest. The years of waiting for bulbs to reach flowering size can seem endless. But if you work out a long-term programme, and sow a limited number of seeds each year, by the end of three years or so the first sowings will be coming into flower.

To start off such a programme, select your bulbs from the chart on this page. These are varieties most likely to flower a few years after sowing. The chart gives basic information about each bulb. But, briefly, here is an outline of the general routine needed to raise bulbs from seed.

Composts Use a loam seed compost for sowing, mixed with sand or grit, if you like, to make it more porous. Use a mix of half loam potting compost and half sand when transplanting. A nursery bed is improved by the addition of leaf mould and sand.

Sowing While some seeds can be sown outdoors, preferably in a cold frame, others will need heat to start them off. Outdoors or indoors, the seeds, which are mostly small, are most conveniently sown in trays or pots. Level the surface of the compost and sow the seed thinly and as evenly as possible. Sprinkle sieved compost lightly over the top and water by placing the container in a basin of shallow tepid water until the compost is moist throughout. Cover the container with a sheet of glass and place in the propagator or cold frame.

Transplanting and after Move the seedlings, as soon as they are big enough to handle, either into deeper trays, spacing them 2in/5cm apart, or into other pots, planting a few seedlings in each. If few seeds have germinated, leave the undisturbed compost in the seed tray for further seedlings to make a belated appearance. About 8 to 12 weeks after transplanting the seedlings, move them to individual pots; the chart indicates when the young bulbs of each species can be planted outdoors.

	SOWING TIME (Germination)	PLANT OUTDOORS	REACHES FLOWERING SIZE
Allium	Spring, 50°F/10°C (6 weeks–1 year)	Late same autumn, or leave in pots	2–2½ years
Alstroemaria	Soak 24 hrs Spring, 65°F/18°C (6 weeks–1 year)	2 years after sowing	4 years
Amaryllis	Soak several hours Spring, 65°F/18°C	4 years after sowing	7–8 years
Brodiaea	Spring, in frame	2 years after sowing	3–4 years
Colchicum	Spring, in frame	In summer, 3 years after sowing	4–5 years
Chionodoxa	Early summer, in frame	Following year, late spring	2–3 years
Camassia	Late spring in frame	Following year, spring	3–4 years
Convallaria	Late summer in greenhouse, unheated propagator	1 year later to frame; 2nd year to outdoors	2–3 years
Crinum	Autumn, 60°F/16°C (2–3 months)	Next summer, mild areas only	2 years
Crocus	Summer, frame (by following spring)	2 years after sowing	4–5 years
Cyclamen	Early winter; outside till spring, then 65°F/18°C	Following year, spring into cold frame for a year, then outdoors	3–4 years
Fritillaria	Summer, in frame or greenhouse (up to 1 year)	Autumn, after first year of flowering	4 years
Galanthus	Summer, in frame	1 year later to nursery bed for 1 further year	4 years
Gladiolus	Early spring, 55°F/13°C	Following spring, then treat as cormlets (p 70)	4 years
Lachenalia	Spring, 65°F/18°C	Grow in greenhouse; pot-on after 1 year	3–4 years
Lilium	Autumn, in cold frame	18 months later	1½–4 years– depends on species
Muscari	Spring, in cold frame	2 years later	4–5 years
Ornithogalum	Spring, in cold frame	18 months later	3 years
Ranunculus	Spring, in cold frame	2 years later	3 years
Scilla	Spring, in cold frame	2½ years later, in autumn	4–5 years

Vegetables

The vegetable plot has a different time scale from the rest of the garden. In planning it you have to think three years ahead, but carrying out the plan each year means a new start on a virtually clean slate. There is also a different orderly discipline in the layout of the vegetable garden, with its rows and beds. The purpose of such planning is to decide where different types of vegetables are to be grown each year. You can then arrange a sequence of manuring, fertilizing and liming to meet the needs of each group of vegetables. Such planning also helps to prevent a build-up of disease, by not growing the same types of vegetable in the same ground each year. This so-called rotation of crops is explained on page 92.

Starting a vegetable garden involves the preliminary hard work of digging, pulling up perennial weeds, and forking in compost or manure for those crops which need it. Ideally, autumn is the time to start. Dig the land, leaving it rough to let frost do the work of crumbling it down to that desirable 'fine tilth'. If you start a fair-sized vegetable garden in spring, it is hard to make up for lost time and do the jobs of sowing, transplanting, weeding and watering as well. Better instead to tackle just part of the vegetable plot; sow and plant that first and leave the rest to be done next year.

BEDS FOR VEGETABLES

Most vegetables are grown from seed sown outdoors. It seems almost a divine ordinance that vegetables should be grown in long, well-separated rows. But until around 150 years ago they were grown in beds, and it has once again been discovered that this is a commercially profitable way of growing them. The same system can be adapted for an ordinary garden,

Vegetable sowing can begin in early spring outdoors, with broad beans among the first to grow. Some gardeners may try to produce even earlier crops by sowing in pots indoors or under glass. But do not transplant to the garden until the soil has had a chance to warm up.

Planting in beds

Growing vegetables in beds instead of rows is an old system revived. Beds measure 4–5ft/1.2–1.5m, with a narrow path between. Close planting results in heavy crops of more tender, and tastier, smaller vegetables.

and the closer planting it involves will produce heavy crops of reasonably sized vegetables, more to modern taste than the once sought-after monster specimens.

Such specimen beds are not more than 5ft/1.5m wide and as long as is convenient. Within them, the seeds are sown in quite close rows *across* the beds, not *along* them. When thinned out, the plants are equidistant in all directions. A path 18in/45cm wide is left between the beds, from which the sowing of seed and all other operations can be done, without having to trample among the plants. This close growing also helps to smother weeds.

Other gardeners may well prefer to stick to the old method of sowing in rows, but even here it is now recognized that the wide spacing recommended in the past was excessive, and in the following advice for individual plants, rather closer spacing is suggested than has been conventional. Generous spacing between some crops is still sensible, however, for instance to minimize the discomfort of picking Brussels sprouts in close-packed rows on a wet winter's day.

Even the most dedicated row gardener might be won over to beds for some crops, whether grown from seed or transplanted from seed beds. These could include summer cabbage, cauliflower, lettuce, leeks and self-blanching celery. The cabbage tribe and leeks are usually sown in seed beds and transplanted. The ground for these should be dug in autumn and given time to settle. When sowing time arrives, the soil will be friable and can be levelled with a rake.

Make the bed no more than 4ft/1.2 m wide, for ease of working, and space the drills 6in/15cm apart, running across the bed rather than lengthways. Having sown each row, label it at once so that you know what has come up; most seedlings look much alike, especially members of the cabbage family.

The methods of sowing, thinning and transplanting hardy annual vegetables are similar to those for hardy annual flowers, already explained in detail on pages 26–27. And vegetable seeds which need to be started indoors are treated in a similar way to that described for half-hardy annuals. If seed is sown under cloches for the sake of an earlier crop, however, the size of a bed or the spacing of the rows will be governed by the width of the cloches.

Asparagus

SEED

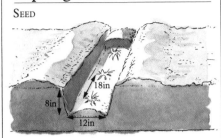

Planting out

Asparagus officinalis, to distinguish it from the ornamental types of asparagus. When grown from seed, allow 4 years before cropping, but an asparagus bed will then last some 10 years. Because of its long life it must be well prepared in the autumn before planting. Dig in generous amounts of rotted manure or compost. In heavy soil, add sharp sand.

Seed is sown in spring. Soak seeds overnight in warmish water and sow in drills 0.5in/1.25cm deep and 12in/30cm apart. When seedlings are 6in/15cm tall, thin to 6in/15cm apart. Keep weed-free and prevent the bed from drying out. The following spring, transplant the asparagus to a permanent bed, spacing plants 18in/45cm apart, in a trench 8in/20cm deep and 12in/30cm wide; allow 3ft/90cm between trenches. Put about 3in/7.5cm of soil back in the trench and gradually fill with the rest of the soil during the summer. In the fourth season, harvest lightly, cutting the spears below soil level when 4in/10cm show above ground.

Aubergine/Eggplant

SEED

Solanum melongena, known as eggplants, because of their shape. These plants can do well out of doors in

mild areas if they have plenty of hot sun in summer, and a sheltered spot; a south-facing wall is ideal. Soil rich in rotted manure is also necessary. Without these conditions they should be grown in a greenhouse, and those grown outdoors need to be covered with cloches in the early stages of planting out. Fruits are picked in late summer, when about 6in/15cm long. Always pick before the shiny bloom fades, or they are likely to be bitter.

Seed is sown in late winter in a greenhouse, in pots of loam seed compost. Place in a propagator at 60–65°F/16–18°C. Germination takes three weeks. When seedlings can be handled easily, pot-on to larger containers, ending with a single plant in an 8in/20cm pot. For outdoor plants, sow seed in early spring and pot-on similarly. Harden off for planting out in late spring. Warm the soil by placing a cloche over the planting area for a week or so beforehand. Plant out 24in/60cm apart. Pinch out growing tips when plants are 6in/15cm tall. When 4 fruits have set, pinch out the remaining flowers and all the side shoots.

Beans
Broad, French, Runner

SEED

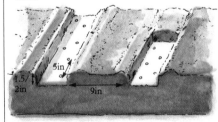

Broad

The broad bean (*Vicia faba*) is the ancient bean of the Old World, while French beans (*Phaseolus vulgaris*) and runner beans (*Phaseolus coccineus*) are the ancient beans of the New World. Most gardeners grow broad beans for the beans and runner beans for the pods. French beans provide both pods and the beans — haricots — inside them. Beans flourish in a rich, well-draining soil, and like sun.

Seed (Broad Bean) is sown according to variety. The hardiest types are Longpods

and used for autumn sowings, but the sweeter-tasting beans are Windsors. Sow these in early spring, 1.5–2in/3.75–5cm deep and 5in/12.5cm apart, in a twin row, separated by 9in/22.5cm. Sow a few extra beans at the end of the row to fill any gaps. Germination takes 1–2 weeks. When the flowers are setting well, pinch out growing tips with 4in/10cm of stem, to encourage an early crop and discourage blackfly. Harvest when young.

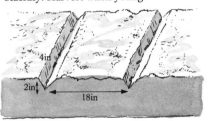

Seed (French Bean) is sown under cloches in spring, removing them in late spring, or in the open from late spring. Sow in drills 2in/5cm deep, and space the beans 4in/10cm apart, with 18in/45cm between rows. Pick when the beans are nearly 4in/10cm long; the younger they are the better they taste, and early picking prolongs the cropping period.

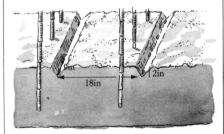

Seed (Runner Bean) is sown in late spring where plants are to grow. Sow 2in/5cm deep, spaced at 9in/22.5cm, in a twin row 18in/45cm apart. Germination takes 1–2 weeks. Put in the supporting canes when 2 leaves have opened; you can then see where the canes should go with little disturbance to the plants. Harvesting begins about 3 months after sowing. Runner beans produce a lot of beans which are often left on the plant too long; not only do they become coarse, but this halts the plant from producing more. The main drawback to runner beans is their need for staking and space.

Vegetables

Beetroot

SEED

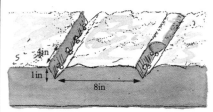

Early crop

Globe varieties of *Beta vulgaris* are for eating freshly gathered during summer and autumn; long-rooted and tankard varieties are good for storing for winter. Deep sandy soil gives the best results for long-rooted beet; no varieties should be grown on newly manured ground or the roots may fork. Choose a sunny spot.

Seed of globe varieties is sown from early spring under cloches, and then every 4 weeks in the open from spring to summer. This ensures crops through summer to autumn. Long-rooting and tankard varieties are sown in late spring or early summer. Beet 'seeds' are in fact fruits, or clusters of seed. To help germination, soak them overnight before sowing. Make drills 1in/2.5cm deep and 8in/20cm apart. The seed clusters are large, so space them out when sowing to cut down thinning later. For early crops, put the clusters 4in/10cm apart in rows; for main crop sowings, 8in/20cm apart. To thin, remove the weaker seedlings from each cluster, leaving one strong seedling.

Broccoli/Calabrese

SEED

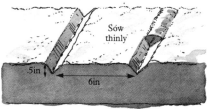

Sprouting broccoli

These are different varieties of the same species of brassicas. Calabrese is now the more fashionable; sprouting broccoli, purple and white, is hardier. Dig the plot in autumn to let the soil settle. In acid soil, scatter lime in winter.

Seed of calabrese is sown at 3–4 week intervals from spring to early summer, for cropping from late summer to autumn. Sow seeds where they are to grow, in drills 0.5in/1.25cm deep. Germination takes 1–2 weeks. Distances between plants and rows can vary enormously, but calabrese do well with less room than often suggested, 20×24in/50×60cm. Close planting will produce small terminal spears around the same time, which is especially useful if you want to freeze them. Wider spacing encourages side shoots over a longer period, and these heads have a better flavour. A reasonable compromise might be to plant 10in/25cm apart, in rows 14in/35cm apart, or experiment on variations allowing at least 12in/30cm square for 2 plants. Cutting begins about 3 months from sowing.

Seed of sprouting broccoli is sown thinly in a seed bed in spring. Transplant seedlings in late spring and early summer when they are about 3in/7.5cm tall. Allow 24in/60cm between plants and 30in/75cm between rows. Plant firmly. Sprouting lasts from late winter to late spring (as long as the central head and side shoots are cut before flowering — when, in any case they are not fit to eat).

Brussels Sprout

SEED

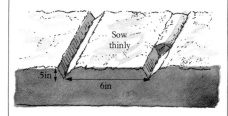

Brassica oleracea gemmifera makes an excellent winter vegetable if the sprouts are picked before they become blousy. F_1 hybrid seeds are the best choice; by selecting different varieties you can have sprouts from early autumn to early spring. Like other brassicas, sprouts need to be grown in firm soil, in part to anchor them against the wind, and they need lime. Do not grow in newly manured soil.

Seed is usually sown in a seed bed and transplanted. Sow an early variety in late winter in a frame, or outdoors in a seed bed in early spring or spring. Mid-season and late varieties can also be sown in spring. Sow thinly in drills 0.5in/1.25cm deep and 6in/15cm apart. Germination takes 1–2 weeks. Thin to not less than 2in/5cm apart when 1in/2.5cm high. Transplant when seedlings are 4–6in/10–15cm high. If you intend to pick sprouts as they reach the size you want, from various plants, allow 30in/75cm between plants and between rows. This encourages extended cropping and gives room for picking. If you want a batch of small sprouts to freeze at the same time, plant more closely — up to 20in/50cm each way. With close spacing, many small sprouts mature at the same time and each plant can then be totally stripped.

Cabbage

SEED

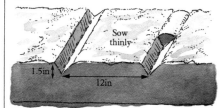

Spring

Brassica oleracea capitata is an all-year-round vegetable which, in the crop rotation method, is usually grown where peas and beans grew the year before. Add no manure. Dig a bed as far ahead as possible, giving the soil time to settle; firmness is important. So also is lime; after digging, scatter some lime on the surface of the soil.

Seed of spring varieties is sown thinly in a seed bed in late summer to early autumn, between 0.75–1in/2–2.5cm deep. Germination takes up to 10 days. Transplant in early autumn or autumn to where they are to grow, in rows 12in/30cm apart. Cabbages should be 4in/10cm apart if they are to be cropped at the spring green stage, or 12in/30cm apart if left to heart. A combination of both spacings may also be considered. Crops are ready to eat in spring.

Seed of summer varieties is sown in a seed bed in spring, 0.5–1in/1.25–2.5cm deep. Germination takes 7–10 days. Thin to 2in/5cm as soon as large enough to handle. Transplant when nearly 4in/10cm in height, planting firmly, 14in/35cm apart, with 14in/35cm between the rows. This should give reasonably sized heads. Vegetables are ready for eating by late summer and early autumn.

Seed of autumn and winter varieties is sown in spring or late spring, 0.5–1in/1.25–2.5cm deep. In summer, plant the seedlings 14in/35cm apart in rows 14in/35cm apart or, for larger heads, 18in×18in/45cm×45cm. Autumn varieties are ready by early autumn and winter varieties from late autumn.

Seed of Savoy cabbages is sown in late spring. These vegetables have a longer cropping period than winter cabbages, but are grown in the same way. Plant out in summer or late summer, 24in/60cm apart each way.

Seed of red cabbage is sown in early autumn in a prepared firm seed bed. Sow in drills 0.5in–1in/1.25–2.5cm deep and thin to 1in/2.5cm apart. In the following spring, plant 24in/60cm each way.

Cabbage, Chinese

SEED

Brassica cernua, Chinese cabbage, produces fast-growing, crisp leaves, to eat raw or stir fried. Some varieties are less liable to bolt than others and are preferred for that reason. They need a humus-rich soil and in a slightly shaded spot may be less liable to run to seed.

Seed is sown where plants are to grow from early summer to summer, a few at a time at intervals. Sow 0.5in/1.25cm deep and 4in/10cm apart; the seeds are quite large and it is worth spacing them out. Thin to 8–10in/20–25cm apart. Keep watered. Tying the leaves together with raffia as they heart will make them crisper. Use 9 weeks after sowing, if they have not bolted.

Carrot

SEED

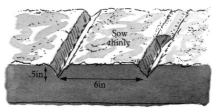

Daucus carota, carrots, grow in all shapes and sizes, but basically there are short-rooted varieties for early crops; medium sized roots pulled when young or left to mature for storing; and long-rooted varieties, best avoided unless you have a deep sandy loam soil. Do not grow carrots in soil recently manured, or the roots may fork, and avoid too much shade.

Seed is sown first in early spring, under cloches, in 0.5in/1.25cm drills which need be no more than 6in/15cm apart. Regular sowings in the open can follow at 3–4 weekly intervals from early spring to the start of summer, to provide a constant supply of young carrots. The later sowings are left to mature for storing (and, if well covered against frost, can be left stored in the ground). Germination takes about 2½ weeks, but is variable. Thin to 2in/5cm apart for cropping when young; 4in/10cm for crops left to mature. Thinning is a dangerous time for carrots, since the smell of disturbed seedlings attracts carrot fly. It helps to thin the plants on a cloudy evening, when there are fewer flies around, to water the bed afterwards and not to leave any thinnings lying around.

Cauliflower

SEED

Brassica oleracea botrytis is not the easiest vegetable to grow, especially in colder areas. It is probably sensible to grow only summer and perhaps some autumn varieties. (Gaps in cropping can be filled by growing mini-cauliflowers for freezing). Prepare the bed where they are to grow in the previous autumn, allowing it to settle. Dig in compost or manure and, if the soil is acid, top dress in the winter with lime.

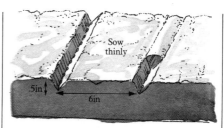

Seed of summer varieties is sown outdoors in spring and transplanted in early summer. These will produce heads in late summer and early autumn. Autumn varieties sown in spring to late spring will provide heads from autumn to late autumn. For worthwhile results the soil must be good, the seedlings transplanted with great care and the plants must never go short of water. Sow seed 0.5in/1.25cm deep, very thinly, in drills 6in/15cm apart. Germination takes 1–2 weeks. Thin seedlings to 3in/7.5cm. Transplant seedlings when they are 4–6in/10–15cm tall. Remove them with as much soil as possible round the roots and replant firmly. Failure to do this may make the cauliflowers head prematurely, producing small useless 'buttons'. Space them 24in/60cm both ways, but with some of the newer Australian varieties, 20in/50cm each way should be enough.

Seed of mini-cauliflowers for eating fresh should be sown successively in small batches. Vegetables for freezing may require one large sowing. Very close planting will produce curds of up to 3in/7.5cm in diameter, enough for a single helping and suitable for freezing. Choose the varieties Predominant or Garant, or others bred for this method of growing. It is better to sow where the plants are to grow rather than to transplant, with probable setbacks. Space them 4in/10cm apart in rows 9in/22.5cm apart.

Vegetables

Celery

SEED

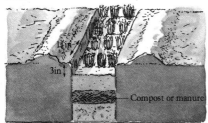

Trench celery

Self-blanching celery (*Apium graveolens*) may have a less pronounced flavour than traditional trench celery, but at least it is less stringy and far easier to grow. Unlike the blanching varieties it is not winter hardy, and is grown to eat from late summer through autumn. Dig in plenty of compost or rotted manure when preparing the outdoor bed in spring.

Seed is sown in early spring or spring, thinly in trays of loam seed compost. Do not cover the seeds. Warmth is needed for germination, most successfully in the range 50–65°F/10–18°C. Keep the compost moist. Germination takes 2–3 weeks. Prick out the seedlings into boxes of loam potting compost, 2in/5cm apart. Gradually harden them off before planting out in late spring. Self-blanching types are planted in blocks, not rows, so that by their proximity they blanch each other. For effective blanching they should be 11in/27.5cm each way at most. A spacing of 6in/15cm each way is the closest to be attempted, producing more, but slenderer, sticks. Plants exposed on the edges of the block will not be totally self blanching. Overcome this by packing straw round the sides of the block or wrapping round a strip of black plastic.

Chicory

SEED

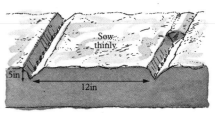

Forced

Cichorium intybus, chicory, comes in two

FLUID SOWING

Unless seeds are large enough to handle separately, it is difficult to space them out properly when sowing. Uneven, bunched sowing leads to overcrowding and poor germination, and makes thinning more difficult. Very fine seed can be mixed with sand, which is some help but not a real solution. An unsteady hand and lack of confidence can still cause small seed to flow unsteadily, particularly in the face of a stiff breeze, which plays havoc with attempts to sow fine seed out of doors.

A moderately simple answer to these problems is fluid sowing. Using this method, seeds are suspended in paste and squeezed along the drills in the seed bed. Research into fluid sowing has concentrated mainly on vegetables, and shows that not only does it make sowing more reliable, but also leads to faster germination, especially in the early, colder months of the year. The explanation is simple. Soil temperature is one vital factor in germination: with most seed, any reluctance to germinate is caused by too low soil temperatures (although very high soil temperatures can also delay germination). In fluid sowing, seed is pre-germinated indoors and then sown outside. Such seed will therefore be growing days and even weeks earlier than seed from early sowings in the open ground. Among vegetables which the system benefits particularly are carrots, celery, onions and, above all, outdoor tomatoes. Later sowings, made when the soil in the garden has warmed up, show some, but not such remarkable, improvement. For a do-it-yourself version of fluid sowing you will need:

Steps in fluid sowing: germinating seed (top left); mixing with flour paste (left); squeezing from bag (above).

A shallow plastic box with a lid
Paper tissues and paper towels
Fine kitchen sieve
Bowl and wooden spoon
Cellulose wallpaper paste, *without* fungicide
A plastic bag
0.25in/0.5cm nozzle
Length of string

varieties: one like slightly tart lettuce; the other the source of chicons used for salads in winter. Chicons, having virtually to be grown twice, take up more time. Because of its long roots, chicory needs a well-dug, as well as rich, soil.

Seed of forcing varieties (such as Witloof) is sown in early summer in drills 0.5in/1.25cm deep and 12in/30cm apart. Germination takes 1–2 weeks. Thin seedlings to 9in/22.5cm; keep weed-free and moist. In autumn to late autumn, when the leaves have died down, carefully lift the tapering roots, cut off the dead leaves 1in/2.5cm above the crown of the root and trim off the root tip to leave it some 9in/22.5cm long. Pack them horizontally in a box of dry sand. Remove a few at a time for forcing — no more than

you are likely to need for a single meal — and plant them in a 9in/22.5cm pot of moist peat or compost, with the crowns of the roots just exposed. Cover the pot with another pot of the same size to block out the light; also plug up the drainage hole. Keep between 50–60°F/10–16°C and in 3–4 weeks the chicons, then about 6in/15cm tall, will be ready for harvesting. If the chicon is carefully broken away from the crown, instead of being cut off, more small chicons will grow. Chicons should be eaten as soon as cut or they will lose their crispness. Seed of non-forcing varieties (such as Sugar Loaf) is sown in early summer to summer. Germination takes 1–2 weeks. Thin to 16in/40cm apart. The vegetable is ready for eating in autumn.

Cucumber/Gherkin

SEED

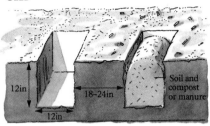

Cucumis sativus. Outdoor (or ridge) cucumbers have developed well in recent years and the so-called Japanese varieties can rival the greenhouse (or frame) types. The typical ridge cucumber was short, fat and knobbly; the newer varieties are long, straight and slim. Gherkins remain short and warty, but many new varieties have been evolved. All need sun and rich soil.

Seed is sown in very late spring or early summer. Prepare in spring by digging holes in a sunny part of the garden, out of the wind, about 12in/30cm wide and 12in/30cm deep, and 18–24in/45–60cm apart. Fill them with a mixture of soil and compost or well rotted manure. Seed can be sown directly in these pockets. Cucumbers will not germinate well if soil temperature is below 55°F/13°C. The plants can be raised indoors in heat but they resent root disturbance, so use peat pots filled with peat seed compost. In spring, plant one seed edgeways, 0.5in/1.25cm deep, in each block. Place in a propagator at a temperature of 70°F/21°C until they germinate. Harden off gradually in readiness for planting outdoors in early summer.

If sowing outdoors, place 3 or 4 seeds in each pocket, at a depth of 1in/2.5cm. After germination, keep only the strongest seedling. After 6 or 7 leaves have grown, pinch out the growing tip so that fruit-bearing laterals will emerge. The Japanese climbing varieties should be trained up netting, and can grow to 6ft/1.8m.

Gherkins are grown in a similar way, but allow only 24in/60cm between plants, which can be left to trail. They are best picked when 3–4in/7.5–10cm long.

Eggplant

SEE PAGE 83

The operation is carried out in four main stages.

1 Sprouting the seed. Spread several layers of paper tissues over the base of the plastic box. Wet the paper, pouring off any excess water. Although the tissues absorb a lot of water, they will go soggy; therefore spread a layer of kitchen paper towel on top. Dampen it. Spread the seeds as evenly as possible over the paper, but not thickly. Put on the lid and keep the box in a warm place. An airing cupboard is not suitable for this, as it will be too warm. The optimum temperature for most common vegetables is 70°F/21°C.

Some vegetables begin to produce roots in two or three days, maybe less, but others will take a week or more. Keep a sharp eye on them, because the next stage must be tackled as soon as germination has begun.

2 Make up the wallpaper paste, following the manufacturer's instructions, but note that it needs to be about *half* the strength required for papering a wall. **NB** *The paste must be a type with no fungicide in it.*

3 When the paste is ready, very gently rinse the germinating seed off the paper into a fine kitchen sieve. Then scatter it into the bowl of paste and stir, most gently, for the growing seeds are easily damaged.

Cut the corner off a plastic bag and

fit it with a nozzle of 0.25in/0.5cm diameter, so that a stream of paste flows easily through. Pour the jelly-like paste into the bag and tie the top lightly with string. You are now ready to sow.

4 Sowing. Take out a drill on the prepared seed bed outdoors, very slightly deeper than for ordinary seed. Water it — this is most important. Then squeeze the paste — and seeds — along the drill and cover with soil so that the paste does not dry out. Water, as well as warmth, is a vital factor in the growth of a seedling. Warmth having successfully germinated the seed, the seed bed must thereafter be kept moist so that the seedlings will grow.

Maybe a word of warning is needed. Although fluid sowing helps to produce earlier crops, do not be tempted to germinate seed before the weather outside is warm enough for the seedlings to grow in the open garden, even under the protection of cloches.

Flower, as well as vegetable, seeds can be sown in this way, but the advantages of a long growing season are fewer for annual flowers and even spacing not so vital. If seeds are listed as needing light for germination, this also applies to pre-germinated seeds. In such cases, leave the lid off the box, but make sure that the paper under the seeds does not dry out.

Vegetables

Leek

SEED

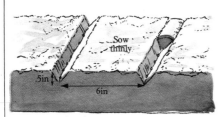

Allium porrum. Although easier to grow than onions, leeks do need some attention. The instructions here are for growing vegetables to eat, not show. Leeks are not too fussy about soil as long as it is reasonably well draining. Dig in compost or rotted manure the autumn before planting.

Seed is sown outdoors in a seed bed in early spring or spring, a little more than 0.5in/1.25cm deep. Germination takes 2–3 weeks. Early summer is the time to transplant the seedlings, when they are 6–8in/15–20cm tall. Lift them very carefully and plant 9in/22.5cm apart in rows 12in/30cm apart. This spacing will give an optimum crop of average sized leeks; closer planting in the row will give slenderer leeks. The method of planting is unusual. Using a large dibber, make holes 6in/15cm deep at the preferred distance along the row and lower a leek into each hole. Then gently pour in a little water; this will give the seedling all the anchorage it needs. Do not try to firm soil round it in the hole. The 12in/30cm space between rows allows room for drawing soil around the stems at intervals from late summer, to make the white part of the stems longer. Lift these very hardy vegetables carefully with a fork, as needed, during the winter.

SPAGHETTI SQUASH

Vegetable spaghetti is a squash. When cooked and cut in half, the flesh inside looks like spaghetti, even though it may not taste like it.

To grow your spaghetti squash, sow seeds in an unheated greenhouse or frame in spring or late spring. Squash resent having their roots disturbed by transplanting, so put your seeds in peat blocks or pots, two at a time. After germination, remove the weaker seedling. In early summer, after hardening off, transplant outdoors in very rich soil. The vegetable spaghetti squash is a trailer, so allow 3–4ft/90–120cm between plants.

As well as being voracious feeders, squashes are incurably thirsty. They are also slug prone, but otherwise should present no problems. These squashes are a rich golden yellow, oval shaped and about 8in/20cm long. Pick when ripe.

To cook the squash, boil whole for 20–30 minutes. Cut in half, remove the seeds and scoop out the flesh. To preserve the illusion that you have 'grown' spaghetti, serve with tomato sauce; or serve cold with mayonnaise.

Lettuce

SEED

Crispheads

Varieties of lettuce (*Lactuca sativa*) fall into 4 main groups: cos, butterhead, crisphead and leaf lettuce. Apart from the risk of bolting in hot weather, there are few problems in growing lettuce outdoors to eat from summer to early autumn. You will need cloches or frames to produce them from early spring to late autumn. And if you want them from winter to early spring, you need a greenhouse or frame with a little heat. Outdoor lettuces need a fertile and not acid soil.

Seed for early summer crops is sown under glass in late winter. Allow 2 seeds to a peat block, removing the weaker seedling. Harden off and plant out under cloches in early spring. The distances apart depend on the variety: dwarfs should be 6in/15cm apart; butterheads in general 9in/22.5cm apart; and the crispheads 12in/30cm.

Outdoor sowings of summer and autumn varieties start in spring and continue into summer, to pick from summer to autumn. Sow in drills 0.25in/0.5cm deep and 12in/30cm apart. Germination takes 1–2 weeks. Thin first to 3in/7.5cm and later to the spacings recommended above for different varieties.

Seed of winter lettuce (a forcing variety) is sown during early autumn or autumn in a greenhouse or heated frame where the temperature remains above 45°F/7°C through the winter months. Sow 2–3 seeds in a peat block and transplant to a bed in the greenhouse or frame, 9in/22.5cm apart. Lettuces should be ready from winter to early spring.

Seed of leaf lettuce is sown between spring and late summer. These varieties never form true hearts but provide a constant source of leaves. The reason is simply that they are planted very closely. The most popular variety is Salad Bowl, with curly leaves, but some cos lettuces can also be used. Sow in rows, 5in/12.5cm apart, at various intervals. Reckon that lettuces from early sowings will be ready in 7 weeks, while later sowings will be ready in under 6 weeks. Aim at sowing 12 seeds, or a few more, to every 12in/30cm of the row. Do not thin, and do not sow too large an area at any one time; 36in/90cm square will be quite enough. When ready, pick the leaves, leaving the stumps in the ground to produce another crop.

Marrow/Courgette

SEED

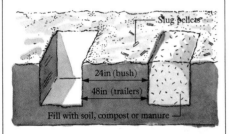

Cucurbita pepo. Many varieties of marrow courgettes (known as summer squash in the United States) are available. Courgettes are infant marrows, but varieties have been bred to bear large numbers of small fruits. Sowing and raising both marrows and courgettes is basically the same. They need rich soil and sun. Dig in compost or manure where the plants are to grow, but not too deeply for the roots do not reach far down. The compost or manure need not be spread over the whole bed – concentrate it where the plants will be.

Seed is sown in early summer outdoors, or in spring indoors in a temperature of 65°F/18°C. Marrows and courgettes do not germinate satisfactorily if the soil temperature is below 55°F/13°C. They also resent root disturbance, so use peat blocks or peat pots and seed compost for planting. Soak the seeds overnight before sowing and put 2 seeds, 1in/2.5cm deep and edgeways (not flat) in each pot. Germination takes a week or a little more. Remove the weaker seedling if both

germinate. Harden off young plants sown indoors before planting out at the start of early summer, in their peat blocks or pots. Allow 24in/60cm between bush varieties and up to 4ft/1.2m for the trailers. Water around the plants, not over them, or the fruits may rot.

Onion/Shallot

SEEDS; SETS; OFFSETS

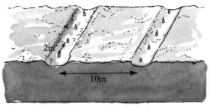

Sets

Onions (*Allium cepa*) can be grown either from seed or from sets, which are immature bulbs grown the previous year. They need a fertile soil and a sunny position to ripen well. It is essential that the ground should be dug well in advance to let the soil settle; they need a firm bed. Spring sown crops (and seed will germinate even at 45°F/7°C) are ready to eat from summer to autumn. Late summer sown crops will be ready in early summer or summer the following year.

Seed is sown outdoors in early spring or spring, depending on local weather, where plants are to grow, 0.5in/1.25cm deep in drills 9in/22.5cm apart. Germination may take 3 weeks. Thin first to 1in/2.5cm or so and later to 3in/7.5cm, for reasonably sized bulbs. As with carrots, thinning can be a dangerous time since the smell of disturbed seedlings attracts the onion fly. Do not leave discarded seedlings lying about. Keep weeding, and do it by hand to avoid damaging the bulbs. Cropping begins in late summer or early autumn. Onions for storing must first be dried off, outdoors in the sun if possible.

In cold areas, seed can be sown in winter in heat (60°F/16°C for good germination). Use peat blocks and sow 2 or 3 seeds, saving only the strongest seedling. Harden off in early spring for planting out in spring — 3in/7.5cm apart in rows 9in/22.5cm apart.

For the earliest crops, sow seed of the

Japanese varieties in late summer. Sow 1in/2.5cm apart in drills 0.5in/1.25cm deep and 9in/22.5cm apart. Do not thin until early spring or spring the following year; then space 3in/7.5cm apart. (Losses during the winter are to be expected.) The crop will be ready in early summer, but these Japanese varieties will not store.

Sets are planted in early spring or spring; they are small onions grown the previous year and stored under heat treatment to prevent the development of flower buds. They finish their cycle of growth when planted the following year. Though more expensive than seed, sets have certain advantages. They are useful in cold areas which have a short growing season, and often give higher yields than onions grown direct from seed. Before planting, first cut off any long, dead tips from the top of the bulbs — birds are tempted to tug at them — but avoid cutting into the flesh. Then plant by pushing the bulbs into the soil, leaving the tips just above the surface. If spaced in rows 10in/25cm apart, with the bulbs 2in/5cm apart, they will produce a heavy crop of medium sized bulbs. Spacing the bulbs 4in/10cm apart in the row gives a smaller crop of larger bulbs.

Shallots

Allium ascalonicum, or shallot, is a milder version of the onion, grown from offsets, which produce 6–12 new bulbs. Plant in firm soil, so prepare the bed well in advance. Choose ground which had been manured for a previous crop, and a sunny part of the garden.

Offsets are planted early, in late winter if possible. Allow 8in/20cm between rows and 6in/15cm between bulbs, pushing them into the soil so that just the tips show above ground. If frost loosens the bulbs, firm back the soil around them. Keep weeded, preferably by hand, for damaged bulbs will rot. The shallots will be ready to harvest and dry off by summer.

Vegetables

Onion
Salad, Pickling, Tree

SEED; BULBLETS

Salad

As well as the so-called dry bulb onions, others are worth growing for use in salads, or for pickling. An oddball member of the group even grows its bulbs in the air, as well as in the soil.

Seed is sown in late summer or early autumn, under cloches, for an early crop of salad onions, alias spring or bunching onions. Sow fairly thickly in drills 0.5in/1.25cm deep and 10in/25cm apart. Make further sowings in the open garden at intervals of 3–5 weeks from early spring to the end of summer. Pull when 6in/15cm high.

A few special varieties of onion are suitable for pickling. These do not need rich soil. Sow the seeds in spring, either scattered over the bed and covered with soil, or in drills 0.5in/1.25cm deep and 6–8in/15–20cm apart. Little thinning is needed. The onions are ready to pull and pickle in summer.

Bulblets of the perennial tree onion (or Egyptian onion, *Allium cepa aggregatum*) are planted in spring or autumn, 12–18in/30–45cm apart, in a sunny, well draining part of the garden. The plant grows 2–3ft/60–90cm tall and produces small but pungent bulblets at the end of stems in place of flowers. In the first year the plant will not produce many bulblets, but will make up for that in the years following. Pick the small bulbs (which are good pickled) before they fall to the ground and take root.

Parsnip

SEED

Pastinaca sativa, parsnip, is not a hard vegetable to grow, but choose the small rooted varieties which need less depth of soil. Grow where the ground has been well manured for a previous crop.

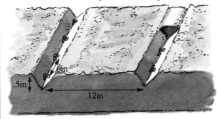

Seed is sown in late winter or early spring, where the plants are to grow (they do not transplant satisfactorily), 0.5in/1.25cm deep, placing 3 seeds together at intervals of 6in/15cm in the rows, which should be 12in/30cm apart. This is adequate spacing for the short rooted varieties. The seeds must be fresh, not any left over from a previous year's packet. Germination is slow, taking 2–4 weeks. Thin out by leaving only the strongest seedling from each group. Take care when weeding not to damage the roots. From autumn onwards, dig up the parsnips as needed, but in cold areas lift the crop in late autumn and store.

Pea

SEED

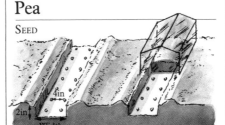

Dwarf

Garden peas (*Pisum sativum*) are either round seeded, which are hardier, or wrinkled, which are sweeter. *Petits pois* are sweeter still, and so are *mangetout*, eaten pods and all. All need rich loamy soil for success and liming is important.

Seed of a round seeded dwarf variety is sown under cloches in late autumn in warm areas for a late spring or early summer crop. Take out a trench 2in/5cm deep and the width of a spade. Put cloches over the trench to warm up the soil before sowing. Peas hate a cold soil. Sow the seeds 2in/5cm apart in lines 4in/10cm apart, along the bottom of the trench. Cover with the soil taken out of the trench and put the cloches back in place. This sowing should be ready for picking in late spring or early summer.

For a midsummer crop, sow a round seeded variety in the open garden in early spring, or wait a little longer until spring to sow an early wrinkled type. Sow in a trench, as above. The general rule is that with more than one row of peas, the rows should be almost as far apart as the expected height of the plants; this may be anything from 18in/45cm to 6ft/1.8m. It is obviously best to choose a variety with a reasonable height in relation to the space in the garden, but keep in mind that taller plants crop more heavily.

For a late summer crop in cool areas, sow a maincrop wrinkled variety in late, spring, sowing as for the earlier crops.

For early autumn and autumn crops, sow a first early wrinkled variety in early summer or summer. When planting at that time of the year, choose a mildew resistant variety.

Seed of *petits pois* and *mangetout*/sugar peas are sown in spring or late spring, when the soil has warmed up. These varieties grow in the same ways as garden peas, reaching about 2ft/60cm. *Mangetout*/sugar pea pods are picked when the peas inside have only just begun to develop.

Potato

TUBER (SEED POTATO)

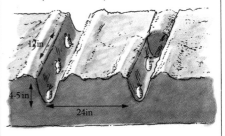

Potatoes (*Solanum tuberosum*) are quite easy to grow, but whether most modern varieties are worth the trouble is debatable. They also take up a lot of room in the garden and storage space in the winter. Even if you find a supplier with a variety you would like to grow — Jersey Royal, for example — there is no guarantee that they will live up to expectations in your garden. The kind of soil in which a potato grows has an enormous influence on its taste. However, there is a lot to be said for 'new' potatoes, straight out of the ground, if room is available in the garden.

Seed potatoes are planted in early spring. They are tubers which have grown the previous year and start into growth again after a period of dormancy. It would take a whole book to explain the behaviour of a growing potato as unearthed by modern research. By-passing the relevance of the physiological age of the tuber and the concept of day-degrees, it is important to remember that the early varieties should be chitted — sprouted — before planting. Egg boxes are excellent for this, the alternative being wooden trays with a layer of dry peat in the bottom. Arrange the seed potatoes in them, rose end up. The rose end is where most eyes are to be found. Sprouting of the first early varieties can be started in winter or late winter, about 6 weeks before it will be safe to plant potatoes outdoors without danger of frost. This may be not until early spring or later, but earlier planting is possible if the rows are covered with cloches or polythene tunnels. Keep the sprouted seed in the light in a cool, frost free room, but out of the sun. By the end of 6 weeks the potatoes will have produced sprouts about 1in/2.5cm long. When it is safe to put them outside, plant them 4–5in/10–12.5cm deep, and 12in/30cm apart, in rows 24in/60cm apart. If there is frost about when the young shoots appear above ground, protect them by covering with a little soil. When the plants have reached 8in/20cm, it is time to earth-up, by drawing up soil against the stems of the plants. The purpose is to prevent the potatoes from greening, which occurs if they are exposed to light. First early potatoes should be ready for digging up in early summer or summer. The sooner they are eaten after harvesting, the better they will taste.

Radish

SEED

Summer salad radishes (*Raphanus sativus*) are either globular or cylindrical. Sow few and often and eat before they grow coarse and hot. Avoid planting on freshly manured ground. Early sowings should be made in a sunny part of the garden, but some shade later will help to prevent the radishes from bolting.

Seed is sown for the earliest crops in late winter or early spring, under cloches. Sow 0.5in/1.25cm deep and space the seeds, which are reasonably large, 1in/2.5cm apart. Germination takes up to a week. Make small sowings in the open every 3–4 weeks from early spring to early autumn, but sowings in hot weather will be the least successful. Depending on the season, radishes will be ready to eat 3–6 weeks after sowing.

Spinach

SEED

Summer

True spinach (*Spinacia oleracea*) includes two types: summer spinach, with smooth round seeds; and winter spinach, with prickly seeds. New Zealand spinach (*Tetragonia expansa*) is a variety which will tolerate hot weather better than true spinach, but cannot stand frost. Spinach beet (*Beta vulgaris*, alias perpetual spinach, but perpetual for only one year) produces leaves for picking over a long

period. All types of spinach should be picked regularly, thereby encouraging the plant to grow more leaves.

Seed of summer spinach is sown every few weeks, starting in early spring, through to the end of late spring, for use from early summer to autumn. Rich soil and a slightly shaded spot is needed. Sow thinly, 1in/2.5cm deep, in rows 12in/30cm apart. Germination takes 2–3 weeks. Thin to 3in/7.5cm apart as soon as seedlings can be handled and later to 9–12in/22.5cm–30cm apart. Water well in a hot dry period or the plants will bolt.

Seed of winter spinach is sown in late summer or early autumn in a sunny part of the garden. Make drills 1in/2.5cm deep, with 12in/30cm between the rows. Thin to 3in/7.5cm and then to 12in/30cm; the later thinnings can be cooked. Pick from autumn to the following spring. Except in warm areas, the plants may need cloche protection during the winter months.

Seed of New Zealand spinach must not be sown outdoors before all danger of frost is over. Soak seeds overnight before sowing. Space them, slightly less than 1in/2.5cm deep, in groups of 3 at intervals 2ft/60cm apart and 3ft/90cm between rows. Thin to leave only the strongest seedling of each group.

Seed of spinach beet is sown twice; once in spring and again in summer. Sow the seeds 1in/2.5cm deep and 4in/10cm apart in rows 12in/30cm apart. Thin to 12in/30cm. Bolting should be no problem. Pick the leaves regularly before they grow too large.

Summer Squash

SEE MARROW/COURGETTE PAGE 89

Vegetables

Some vegetables grow happily sharing the same diet as others, and so, on dietary grounds, garden vegetables can be split into groups, usually three. These are: *legumes*, *brassicas* and *roots*. For the different vegetables in each group, see the illustrations below.

To cope with the different needs of these groups, divide the vegetable garden into three, not necessarily equal, parts. Peas and beans need a rich soil, so dig plenty of well rotted manure or compost into their bit of the garden in the autumn/winter before planting. Peas and beans, with the help of the soil's micro-organisms, fix nitrogen in the soil from the air and so leave the soil richer in nitrogen than they found it.

This suits the leafy brassicas, when they move next year into the beds left vacant by the peas and beans. To satisfy the brassicas' different diet, their patch is limed (during the winter months) and scattered with fertilizer a few weeks before sowing or planting. No manure need be given unless the soil is in obvious need of it.

In the third year, roots take over the space previously occupied by the brassicas. The ground is given only fertilizer, not manure, which tends to

ROTATING CROPS

make the roots grow misshapen. Nor do roots get any lime. If potatoes are grown alongside the roots, they can profitably be manured, but must not be limed.

If you have room, and intend to grow a lot of potatoes, you can run a four-year instead of a three-year rotation, allowing potatoes to take over from the roots before handing the ground back to the peas and beans. So, year after year, the cycle goes on, as the vegetables change their places in the garden. This is the meaning of the 'rotation of crops'.

KEEPING HEALTHY

Not growing the same plants in the same ground two years running involves more than just the need for manuring or liming. An equally important aim of the system is to limit the spread of disease. Different vegetables are plagued by different diseases and attract different pests. Therefore, growing one type of vegetable in the same ground year after year might cause a catastrophic build-up of disease. But in a rotated system,

the diseases specific to peas, for example, would not harm the following year's cabbages. Brassicas have plenty of diseases peculiarly their own, and these would not be passed on to the roots which follow them. Although rotation does not guarantee a disease-free vegetable garden, it does help to limit the problems.

In practice, the rotation of crops never works out as it does in theory, because the amount of space given to each group of vegetables is never likely to be equal. Inevitably, therefore, parts of some crops will be grown in the same place as the year before. But even if not perfect in execution, the system should be followed as far as possible. It works.

The rotation method is achievable only because most vegetables are grown from seed every year. But the kitchen garden also has its perennials, such as asparagus and rhubarb, and these, together with the soft fruits, must have permanent homes of their own. Even these old faithfuls are not immortal, however, and when past their best, weakened by old age or the declining fertility of the soil they inhabit, another generation should be planted, making a new start in another part of the garden.

Group A Mainly *legumes* — the peas and beans — along with onions, leeks, celery, tomatoes and lettuce.

Group B *Brassicas* — cabbages, Brussels sprouts, cauliflowers, broccoli, spinach; also turnips and swedes, which, although roots, belong to the brassica family.

Group C *Roots* — beet, carrots and parsnips, plus potatoes.

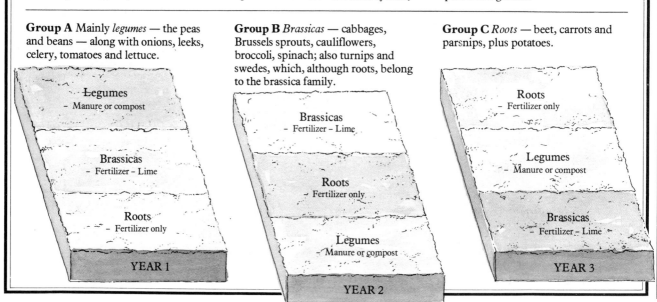

YEAR 1 — Legumes – Manure or compost; Brassicas – Fertilizer – Lime; Roots – Fertilizer only

YEAR 2 — Brassicas – Fertilizer – Lime; Roots – Fertilizer only; Legumes – Manure or compost

YEAR 3 — Roots – Fertilizer only; Legumes – Manure or compost; Brassicas – Fertilizer – Lime

Sweet Corn

SEED

Late spring or early summer planting

The taste of sweet corn (*Zea mays*) picked from the garden is totally different from the often stale shop-bought product. New F₁ hybrids now make it possible to grow sweet corn in both mild and cool areas, but the site must be sunny and the soil well manured. When the silks at the end of the cob begin to wither in late summer, test for ripeness by pressing a seed. If a clear liquid emerges, it is too early; a milky liquid shows it is just right; a thick or dry liquid indicates it is too late.

Seed is sown in spring in peat pots of loam seed compost. Sow 2 seeds to a pot, as sweet corn dislikes transplanting. Place in a propagator at 55°F/13°C. Germination takes 2 weeks. If both seeds germinate, remove the weaker. Plant out in late spring or early summer in blocks, not a single row, 18in/45cm apart. This helps pollination, as the male flowers growing at the top of the plant have to pollinate the silks of the lower growing female cobs. Seed can be sown outdoors in late spring or early summer when all danger of frost is over. Sow 2 seeds together at 18in/45cm intervals in rows 18in/45cm apart. Thin out the weaker of the two.

Tomato, Outdoor

SEED

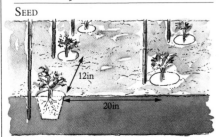

Growing tomatoes (*Lycopersicon lycopersicum*) outdoors is a gamble in cold areas. The plants must be started indoors in heat, but not too early; it is pointless to have young plants ready for planting outdoors when it is too cold for them to be moved there. This waiting for weather can well mean that the crop does not begin to ripen until late summer and then continues for only a few weeks.

That, at least, was the position with the older varieties, together with the business of staking, and pinching out of growing tips and lateral shoots. With some newer varieties, there is a far better chance of a worthwhile crop.

For the earliest crops, choose the so-called sub-arctic varieties of dwarf, or bush, tomatoes, such as Sub-arctic Plenty. The plants grow to about 12in/30cm tall and the tomatoes reach only 1in/2.5cm in diameter.

Seed of sub-arctic varieties is sown outdoors in late spring and thinned to only 12in/30cm apart. The plants must not be pinched out. Fruit sets well even in cold conditions and the crop should start to ripen in 8 weeks. If sown and grown under cloches, or in polythene tunnels, the first fruits could be picked a week earlier. One great virtue of these varieties is that the ripe fruits are slow to deteriorate on the plant. Cropping may well continue until the first frosts.

Even earlier crops can be obtained by pre-germination and fluid sowing (explained on page 86). The germinated seed can be sown under cloches or in the open ground. If using cloches, place these over the planting area at the beginning of spring, to warm up the soil a little. In the middle of spring, pre-germinate the seed indoors. Place 2 or 3 of the pre-germinated seeds at intervals of 12in/30cm in a single row along the cloches or tunnel, and a depth of 0.75-1in/2-2.5cm. The seedlings will take 1-2 weeks to show through. No thinning may be needed.

If planting in the open, wait until the end of spring before putting the pre-germinated seed outdoors. Space them in pairs about 12in/30cm apart, with 20in/50cm between the rows. Apart from keeping down weeds until the plants are large enough to smother them for themselves, there is little to be done except waiting to pick the fruit.

Although this early planting is neither totally foolproof nor frost proof, given a moderately good summer it should prove worth the trouble.

Turnip/Swede

SEED

Turnips (*Brassica rapa*) grow quickly and are picked young; therefore sow a few often. The sweeter, milder swedes (*Brassica rutabaga*) grow to maturity and require only one sowing for the year's supply. Do not sow in freshly manured ground and if the soil is acid, lime it during the winter. Avoid shady spots.

Seed of turnips is sown in 0.5in/1.25cm deep drills in early spring. Follow with sowings every 3–4 weeks to summer. Germination takes 7–10 days. Thin to 3in/7.5cm apart and later to 5in/12.5cm. Turnips are ready for eating in 6–10 weeks, before they grow too big. Avoid sowing too many at any one time or you will produce a surfeit of stringy turnips.

Seed of swedes is sown in late spring in warm areas and in early summer in cold areas. Space the seeds thinly in a drill 0.5in/1.25cm deep. Germination takes 7–10 days. Thin seedlings out in stages until 10in/25cm apart. These are very hardy vegetables and can be left in the ground until required, or they can be lifted and stored in early winter.

Herbs

Vegetables take up a lot of room in the garden and need a lot of attention. Herbs require less space and less time.

From the cook's point of view the ideal place for a herb garden is near the kitchen door. As for the herbs themselves, most would prefer a place in the sun, in a well drained and fertile soil, but one which is not over rich. These adaptable plants seem ready to compromise, however.

Besides tasting good, many herbs also look attractive, having a wide variety of leaf colours and shapes, even though most belong to one of only two plant families: the Labiatæ (Deadnettles) and the Umbelliferæ (Hemlocks). If your aim is to make the herb patch look pleasing, it is wise to sketch a plan before making a start, so that you can sort out where the tall and small plants are to go. But however desirable beauty may be, there is one very important practical point to bear in mind: herbs are grown to be picked and so should be easily accessible, even — or especially — in pouring rain! One solution is to arrange the herb garden in a pattern of flags, so that picking for the kitchen does not involve floundering around in wet soil.

If you want to grow only a few herbs, or if space is limited, some will make themselves quite at home in a flower border, or else can be grown on a patio in tubs or in growing bags. With few exceptions, seeds of herbs, like those of vegetables, are sown outdoors. But even those which must be started indoors in gentle heat are not difficult to cultivate.

In cold areas some of the perennials, such as rosemary, will need protection in winter. Among biennials, too, are some which cannot be expected to survive a hard winter. And the annuals, of course, will die down. To ensure a year-round bounty from your herbs, dry or freeze some of the summer harvest.

Among the herbs which respond readily to division are chives; new clumps from old are obtained with little difficulty. Rosemary cuttings are taken in early spring.

PLANTING HERBS

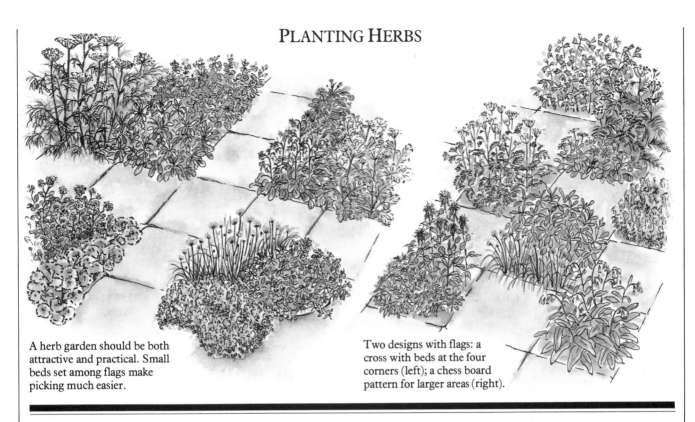

A herb garden should be both attractive and practical. Small beds set among flags make picking much easier.

Two designs with flags: a cross with beds at the four corners (left); a chess board pattern for larger areas (right).

Balm

SEED; DIVISION

Melissa officinalis, Lemon Balm. Perennial, but easily damaged by frost. Lemon scented plant with honey-sweet leaves, attractive to bees. Grows up to 36in/90cm, with a spread of 18in/45cm.

Seed is sown in a cold frame in early spring or spring. Germination takes 2–3 weeks. Plant out in late summer or early autumn, 12in/30cm or more apart. **Division** usually takes place in spring. Divide the root into pieces with 3–4 buds and plant 24in/60cm apart. Will need protection through the winter.

Basil

SEED

Ocimum basilicum, Sweet Basil. Tender annual with a strong clove-like flavour. Grows to 18in/45cm. *Ocimum minimum*, Bush, or Dwarf, Basil. Perennial, but treat as a half-hardy annual and raise from seed each year. Grows 6–12in/15–30cm. Not as strong a flavour as Sweet Basil.

Seed is sown indoors in gentle heat in late winter and early spring. Since basil does not transplant easily, sow 2 or 3 seeds in a peat pot to avoid root disturbance when planting out. Germination takes 10–14 days. Plant out in early summer, 12in/30cm apart for Sweet Basil, and 8in/20cm apart for Bush Basil. Seed can also be sown outdoors in spring to late spring.

Chervil

SEED

Anthriscus cerefolium. Biennial grown as an annual. Its leaves are like a more delicate version of parsley, with a sweet, slightly aniseed, flavour. The plant grows 12–18in/30–45cm.

Seed is sown outdoors at intervals of several weeks between early spring and late summer for a constant supply of tender young leaves. Sow 0.25in/0.5cm deep and thin seedlings to 10in/25cm apart; they cannot stand transplanting. Leaves will be ready for picking from 6 weeks after sowing.

Chives

SEED; DIVISION

Allium schoenoprasum. Hardy perennial forming clumps of narrow, delicately onion flavoured leaves, about 8in/20cm long.

Seed is sown in spring, several to a soil block so that when planted outside they will more rapidly form a clump of bulbs. Plant the blocks about 8in/20cm apart. **Division** takes place in spring or autumn. Take not more than half a dozen bulbs to form each new clump, 8in/20cm apart. It is general practice to divide clumps when the bulbs become too crowded, say every 3 years.

Herbs

Dill
SEED

Anethum graveolens. Hardy annual with a mild aniseed flavour. Grows to 3ft/90cm.

Seed is sown outdoors from spring where plants are to grow, and just covered. Germination takes 2 weeks. Thin to 9in/23cm apart. Make successive sowings until early summer or summer to ensure a constant supply of young leaves. If the plant is grown for its seed, and not just for the leaves, a spring sowing gives it time to mature.

Fennel
SEED; DIVISION

Foeniculum vulgare. Hardy perennial, reaching 5ft/1.5m or more. The herb, grown for its leaves, should not be confused with the vegetable Florence Fennel, grown for its swollen stem base. Both have a pronounced aniseed flavour.

Seed is sown in spring to late spring, where plants are to grow; if in rows space them 18in/45cm apart. Unless the weather is cold, germination takes less than 2 weeks. Thin to 18in/45cm apart.
Division takes place in early spring. Lift well-established roots, split into portions — not too small — and replant 18in/45cm apart.

Garlic
DIVISION

Allium sativum. Hardy perennial. The bulb, divided into cloves, has a distinctive and persistent onion flavour. It is hardier than is often thought.

Division Remove the papery covering of the bulb and split it into cloves, discarding the smallest. Plant in autumn or late winter, 5in/12cm apart, in rows 12in/30cm apart. Make only a shallow drill — 1in/2.5cm plus — so that the necks of the cloves are just showing above the soil.

PRESERVING HERBS FOR THE WINTER

To dry herbs, pick when they are young and tender, before the plants have flowered. Choose the morning of a dry, warm day. Remove any damaged or diseased patches, then tie the herbs in small bunches and wash them. Drain and place on absorbent kitchen paper to dry, but keep them out of the sun, which ruins aroma and colour.

Choose one of three drying methods:

1 Hang the bunches upside down in a dark, well-ventilated room for a week or more;

2 Spread the herbs on a tray and place in warmth and darkness (an airing cupboard, for instance), turning them from time to time. Drying time takes 2–3 days;

3 Spread the herbs over a cloth on a tray in a cool oven, *leaving the door open*. The best temperature is between 70–90°F/21–33°C. Turn over after half an hour. They will dry in about an hour.

Crumble the crisp leaves between your fingers and store in airtight jars in the dark.

Freezing herbs causes loss of crispness necessary for garnishing, but colour and flavour will be better than dried herbs.

Pick herbs for freezing in the morning, wash, and shake the water from them. Tie up small separate bundles of each herb, so that flavours do not mix, and put several into a polythene deep freeze bag. Otherwise, mix a few herbs together, *bouquet garni* style, tie, and pack several in a bag.

A simple alternative is to chop the herbs, fill ice cube trays with them, top up with a little water and freeze. Turn out the cubes when frozen and put into polythene bags. These make excellent flavouring cubes.

Marjoram
SEED; CUTTINGS; DIVISION

Origanum majorana, Sweet Marjoram. Perennial best treated as a half-hardy annual. The plant is sweetly aromatic.

Seed is sown in early spring in a tray of peat seed compost. Keep at 50–55°F/10–13°C. Thin out and in late spring transplant outdoors, spacing 9in/22.5cm apart.

Origanum onites, Pot Marjoram. Hardy perennial, the leaves of which die down in winter. Less well flavoured than *Origanum majorana*.

Seed is sown thinly outdoors in spring. Germination is slow but eventually the seedlings should be thinned to 12in/30cm apart.
Cuttings (basal offshoots) In spring or late spring, take 2in/5cm cuttings of basal offshoots. Remove lower leaves and insert cuttings in pots of half peat and half sand. When rooted, transplant to their permanent site.
Division Established plants, 2 or 3 years old, can be divided in spring or autumn.

Mint

SEED; DIVISION

Mentha spicata, Common Mint, or Spearmint; *Mentha rotundifolia*, Apple Mint, a better taste than Common Mint; *Mentha rotundifolia*, Bowles variety, known as Bowles Mint, is better still; *Mentha crispa*, Curly Mint, strong flavoured; *Mentha citrata*, Eau de Cologne, fragrant lemon mint; *Mentha rotundifolia variegata*, Pineapple Mint, greatly given to spreading. The species of mint are endless, in various strengths and flavours.

Seed is sown outdoors in spring. Transplant to 9in/22.5cm apart.
Division Dig up some roots in spring or autumn and cut the creeping stolons into 4in/10cm lengths. Plant in drills 2in/5cm deep and 9in/22.5cm apart. The problem is not likely to be in starting growth, but in keeping the plants in check later on, given fertile and moist soil. It is best to divide the roots and start another bed in a new position. Dig up plants in early spring and remove stolons 4–6in/10–15cm long, with new shoots. Replant 9in/22.5cm apart.

Parsley

SEED

Petroselinum crispum, Curly Parsley. Biennial best grown as an annual. A most widely used herb, with an aromatic flavour which cannot be likened to any other.

Seed is sown in spring, when the soil has warmed up, in drills 0.5in/1.5cm deep. If in rows, space these 12in/30cm apart. Germination can take 5–8 weeks and during that time the soil must be kept moist. The seedlings are easily harmed by overcrowding, so thin them when 1in/2.5cm tall to 3in/7.5cm apart and later to 8in/20cm. Successive sowings can be made until late summer (although the plants may not survive severe winters.)

Rosemary

SEED; CUTTINGS; LAYERING

Rosmarinus officinalis. Tender shrub needing some protection except in mild winters. An aromatic plant, it produces a pungent flavour, very pervasive in cooking.

Seed is sown outdoors in a shallow drill towards the end of spring. Germination may be very slow. When seedlings are large enough to handle, thin to 2in/5cm apart. Transplant when they are 3–4in/7.5–10cm high, placing them 6in/15cm apart. When moved to their final position in early summer or summer, they should be 3ft/90cm apart.
Cuttings are taken in early spring, or in

CLASSIC CULINARY HERBS

Fines herbes A mixture of finely chopped herbs, usually parsley, chervil, tarragon and chives.
Bouquet garni. Strictly two or three sprigs of parsley (with stalks) to one small stalk of thyme and one bay leaf, tied together. Some cooks add thyme, marjoram and winter savory. Dried herbs are often wrapped in muslin as well.
Mixed herbs Chopped fresh or dried herbs, often parsley, sage, thyme and marjoram.
Buerres composés Knobs of butter mixed with herbs as a garnish. *Beurre maître d'hotel* (with parsley, *buerre aïl* (with garlic), *buerre estragon* (with tarragon).

ripe cuttings, 6in/15cm long. Insert these in pots of half peat and half sand and place in a cold frame or under a cloche. When rooted, transplant into pots of potting compost. Keep under a cloche through the winter months and plant in their permanent site in late spring of the following year.

Alternatively, hardwood cuttings, 6–8in/15–20cm long, can be taken in early autumn or autumn and planted where they are to grow. It is wise to cover these with a cloche through the winter months.
Layering In summer, peg down some of the lower branches into sandy soil. When well rooted, cut away from the parent and transplant.

Sage

SEED; CUTTINGS; LAYERING

Salvia officinalis. Many varieties include the commonly grown Broad-leaved Sage, a perennial shrub; and the more tangy Narrow-leaved Sage, a perennial often treated as an annual. All varieties are strong smelling, with an astringent taste, and grow up to 12–24in/15–30cm.

Seed of Narrow-leaved Sage is sown in early spring in a frame or greenhouse. Use peat seed compost and cover the seed very thinly. Transplant into small individual pots after the first true leaves have appeared. When well established, plant outdoors 12in/30cm apart. Alternatively, sow outdoors in late spring. Germination takes about 14 days. Transplant seedlings to a nursery bed, and in autumn plant out to their permanent site, 12in/30cm apart. Treated then as perennial, they may need some protection in winter.
Cuttings With Broad-leaved Sage (which seldom sets seed) take 3in/7.5cm heel cuttings from early summer to early autumn and plant in a half-peat, half-sand mixture. Place in a frame and plant out in early spring or spring the following year. Otherwise, in early autumn or autumn, plant hardwood cuttings, 6–8in/15–20cm long, where they are to grow. Cover in winter.
Layering In early spring, anchor some stems to the soil. In the late spring, cut away rooted portions from the parent plant.

Salad Burnet

SEED

Sanguisorba minor is an almost evergreen perennial which grows to 12–15in/ 30–40cm. It is an attractive plant and one of the few which provides fresh green leaves during the winter. The plant also readily seeds itself, a virtue because young leaves are more tender than those of the mature plant. The flavour is something like that of cucumber.

Seed is sown outdoors in spring in drills a little less than 1in/2.5cm deep. Germination takes about 3 weeks. Thin the seedlings, but they will not transplant.

Savory

SEED; DIVISION; CUTTINGS; LAYERING

Satureia hortensis

Satureia hortensis, Summer Savory. Hardy annual, growing up to 12in/30cm tall, with peppery tasting leaves. *Satureia montana*, Winter Savory. Hardy evergreen perennial, also growing up to 12in/30cm tall, but more compact.

Seed of Summer Savory is sown outdoors in spring or late spring and covered very lightly. Germination takes 10–14 days. Thin seedlings when 2in/5cm tall to about 6in/15cm apart. Winter Savory needs light to germinate, so its seed should not be covered. It germinates in

much the same time as Summer Savory, but thereafter grows more slowly.
Division Split the roots of old Winter Savory plants in early spring and replant.
Cuttings In late spring, take cuttings of side shoots from Winter Savory and plant in their permanent positions.
Layering In spring, bend a few stems of Winter Savory on to the soil and anchor down. They will root in a few weeks. Then cut away from the parent and plant out where they are to grow.

Sorrel

SEED; DIVISION

Rumex scutatus, French Sorrel. Perennial, easily grown. This acidulous plant grows about 2ft/60cm tall.

Seed is sown outdoors in spring, thinning to 10–15in/25–40cm apart.
Division In spring or autumn, divide the roots and plant 15in/40cm apart.

Sweet Cicely

SEED; DIVISION

Myrrhis odorata. Hardy perennial, a rampant plant with feathery leaves which have a sweet aniseed flavour. Grows 3ft/90cm tall, and far more in time. It dies down in autumn.
Seed is sown outdoors in early spring or spring. When transplanting, space 18in/45cm apart.
Division In spring or autumn, dig up the

tap roots with great care. First chop off the long tapering part. Then cut the rest of the root into pieces which each have only one 'eye'. Bury each piece 12in/30cm apart, with the 'eye' 2in/2.5cm below the surface of the soil.

Tarragon

DIVISION

Artemesia dranunculus. Hardy perennial, one of the classic culinary herbs. Its bitter sweet flavour should be used with discretion. French tarragon has the most flavour, but Russian tarragon is the hardier. They grow to a height of 24in/60cm with a spread of over 12in/30cm, and a far larger root spread when established.

Division takes places in early spring or spring. Dig up the plant and pull the underground runners apart; do not cut them. After dividing the old plant, replant the runners about 3in/7.5cm deep, leaving 16in/40cm between each new plant. It is best to renew plants in this way every 2 or 3 years.

Thyme

SEED; CUTTINGS; DIVSION

Thymus vulgaris, Common Thyme. Hardy perennial which grows 8in/20cm tall. The herb has a strong flavour unlike any other. *Thymus citriodorus*, Lemon Thyme. Hardy perennial, but less hardy than Common Thyme. There is also a golden-green leaved Lemon Thyme with the same lemon scent.

Seed is sown outdoors from spring to late spring, in drills 0.25in/0.5cm deep. Thin to 2in/5cm apart, and finally plant out 12in/30cm apart.
Cuttings From spring to late spring, take cuttings, 2in/5cm long, of new lateral shoots with a heel from the old stem. Transplant in early autumn, 12in/30cm apart.
Division Roots of established plants can be divided in spring or late spring.

Soft Fruit

The ideal kitchen garden would be walled, and spacious enough not only for vegetables but all kinds of fruit as well. The walls would be covered with top fruit, trained as fans, espaliers or cordons, with perhaps a fig. Soft fruits, in this ideal retreat, would ripen to juicy fullness under the protection of wire cages preventing birds from pecking at the buds in winter and berries in summer. There would be a glasshouse for vines and, beyond the garden wall, an orchard. So much for the dream.

In reality, most gardeners have to be content with room for soft fruit only. This is certainly the most rewarding to grow, and most economical in space. Within the limits imposed by space, choice of fruits must first be made among those you like most. If need be, it can then be narrowed down to fruits which should be eaten soon after being picked — such as strawberries with the warmth of the sun still on them, rather than green gooseberries which have a longer off-the-bush life.

Soft fruit of some sort is now available in shops for much of the year, carried all over the world by air ('for freshness') but the season for garden crops is unhappily short. After the blackberries of autumn have gone, there is a long wait for next year's early summer strawberries, along with the gooseberries (unless the gap is bridged by the glorious pink shoots of early forced rhubarb). On the heels of strawberries come raspberries and currants, followed by a few autumn fruiting strawberries. Then on to blackberries and the blackberry hybrids again. The fresh soft fruit season is over for the year, and it's back to the freezer.

Apart from eating the crops, a variety of pleasures are to be gained from propagating the plants. Very few are grown from seed (because they will not breed true) and so gardeners have to depend on asexual methods – such as, hardwood and leaf bud cuttings, simple and elaborate layering, using suckers and runners. There can be no boredom with these tasks.

Strawberries layer themselves from runners, but keep only those from healthy, virus-free plants.

Soft Fruit

Blackberry

TIP LAYERING; LEAF BUD CUTTINGS

Rubus fruticosus. Hardy shrub, needing plenty of room — anything from 6ft/1.8m to 10ft/3m.

Tip layering takes place in late summer. Pick a cane long enough for its tip to be bent over and buried in the ground, and tough enough to be bent over without snapping. In the soil below, make a hole with one side vertical and the other at an angle of 45°. Put a handful of peat mixed with sand into this, and bend the cane over so that 6in/15cm of its tip is in the hole. Cover with soil and put a stone or brick on top to stop the tip from forcing its way out. Leave until spring. If it has rooted, sever the cane about 8in/20cm above the point where it enters the soil; if not, leave unsevered until shoots appear through the soil, or until autumn. Then lift the rooted cane and plant where required. Allow 6ft/1.8m between plants for the less vigorous varieties and 10ft/3m for the more exuberant.

Leaf bud cuttings are taken between midsummer and early autumn. This interesting but troublesome method could be employed for some hybrid berries, which are less successfully propagated by tip layering than are ordinary blackberries. A propagator is needed, and some heat.

Cut a stem into pieces, each with a leaf and a bud. Cuttings should be taken from semi-ripe wood of the present year's growth, and not the soft tip of the stem. Push them 4in/10cm apart into peat compost, with added sand, so that only the leaf shows above the surface. Water the cuttings and keep the propagator closed, probably for up to a month, while roots form. Then gradually increase the amount of ventilation. In 8 weeks or more a root system should have developed. The cuttings can then be transplanted to a nursery bed before being moved to permanent quarters the following year.

Hybrid berries can be propagated in similar ways — among them loganberry *Rubus loganobaccus*, boysenberry and tayberry.

Currants

HARDWOOD CUTTINGS

Ribes nigrum, Blackcurrant. Propagate only from totally healthy bushes. They

should not be suffering from gall mite, or big bud mite (buds are swollen), or from reversion virus, which is suspected if big bud is present and if crops diminish for no obvious reason.

Cuttings are taken from early to late autumn. Choose well-ripened shoots of the present year's growth. Remove the soft (unripe) top tip and then, carefully, the leaves. Do *not* remove any buds. Reduce the cuttings to 9in/22.5cm in length, cutting the top just above a bud and the bottom just below a bud. In the soil, make a trench with one vertical side and line the bottom with a little coarse sand. Insert each cutting, 9in/22.5cm apart, against the vertical side of the trench. Plant deep enough to leave only 2 buds above soil level; all other buds will be buried. In coming years these will provide new shoots to build up the bush. Firm the soil around the cuttings when filling up the trench. Move to permanent positions the following autumn.

Steps in tip layering blackberries: **1** Bend cane to ground. **2** Bury 6in/15cm of cane in shallow hole. **3** Hold it down with a stone. **4** Sever cane from parent when rooted. **5** Lift rooted cane. **6** Replant.

Steps in taking blackcurrant cuttings: **1** Remove soft tip and leaves. **2** Trim to size, but do not remove any buds. **3** Dig narrow trench and plant cuttings, leaving only two buds above soil level. **4** Firm soil around cuttings. **5** Plant out where they are to grow the following autumn.

Ribes sativum, Red Currant and *Ribes grossularia* var. *uvacrispa*, White Currant, are grown with a bare stem (or as cordons).

Cuttings are taken from early to late autumn. Unlike blackcurrants, the lower buds (which if buried would produce additional shoots instead of a single stem) are rubbed off. Do this with your fingers, leaving 3 or possibly 4 buds at the top. For adequate length of stem, the cuttings should be 12–15in/30–37.5cm. In the soil, make a trench with one vertical side, and insert the cuttings about half their length into the ground, firming the soil around them when filling in the trench. If growth has been good, move the plants to their permanent site the following autumn. The shoots which have grown should then be pruned back to an outward-facing bud no higher than 3 or 4 buds from the base of the shoot. Hard pruning is vital at this stage if the plant is to be shaped into a heavy bearing bush.

Take long hardwood cuttings of red and white currants and leave only 3 or 4 buds at the top to make a bush with stem.

Gooseberry

HARDWOOD CUTTINGS

Ribes grossularia. There may be more failures in propagating gooseberry cuttings than with other soft fruits, but the chances of success are certainly improved by dipping the cuttings in hormone rooting powder.

Take hardwood cuttings of this year's growth, with a heel. Remove leaves, thorns and all buds except top 4.

Cuttings are taken in early autumn or autumn. Use cuttings about 12in/30cm long, from the year's new growth, preferably with a heel from the old wood. Remove any leaves and thorns, and all buds except the top 4. Plant 6in/15cm deep in a shallow trench, and firm the soil around the cuttings. There will be little growth the following summer, so wait until the next spring before planting them out.

Soft Fruit

Raspberry

DIVISION OF SUCKERS; SEED

Rubus idaeus. Raspberries have wandering roots and suckers which are thrown up in an indisciplined way between the rows. These are a nuisance, interfering with cultivation and shading well-behaved canes from the sun. Nonetheless they are a ready-made source of new canes, but only if the parent plants show no sign of virus disease. Any disease will be passed on to the new canes.

To start a new row of raspberries, dig up sturdy rooted suckers in autumn. Sever from parent; replant.

FRUIT FOR HEALTH

Berried fruits not only taste good, but *are* good for you, even though they may be between 80 and 90 per cent water. They are rich in vitamins.

The currants and gooseberries are particularly high in vitamin C, or ascorbic acid, the vitamin which helps in building tissue that binds the cells of the body together.

Gooseberries, red currants, loganberries and blackberries are a good source of thiamine (B_1), one of the seven B vitamins essential in our diet; a heavy deficiency of this may lead to nervous disorders. Niacin, or nicotinic acid, is another of the important B vitamins, and is found in raspberries and strawberries. B_2, or riboflavin, is found in currants, raspberries and strawberries.

Carotene, which is converted in the body to make vitamin A, is also to be found in gooseberries.

Soft fruits also contain useful minerals; blackberries, though low in vitamin C, are richer in calcium.

Division of suckers takes place in autumn. With a fork, carefully lift up the strongest suckers, along with some of the root, severing roots cleanly from the parent plant. With these canes start a new row of raspberries, 18in/45cm apart. After planting, cut the new canes down to within 6in/15cm of the soil.

Seed is sown after fruiting. This is always something of a gamble, but may produce a new variety to supersede all others. Pick some berries, still on the stalk, from plants free from virus disease. Keep indoors exposed to the sun until fully ripe and the stem comes away readily. Swish round in a bowl of water to make the seeds come away from the flesh. Drain off the water, remove seeds from the flesh and wash them clean. Spread on a cloth to dry in the sun. When thoroughly dry, sow the seeds in compost, with added sand, and keep in a temperature of 55–60°F/13–16°C. When large enough to handle, prick out into individual pots, until ready to be planted outdoors in spring. With luck they should produce some fruit in their second year.

Rhubarb

DIVISION; SEED

Rheum rhaponticum. Hardy perennial. Propagation is usually by division, but rhubarb can be grown from seed, although it means waiting longer for a crop.

Division takes place from autumn to early spring when the plant is dormant. With a sharp spade, split the central clump of the plant into portions, ensuring that each has at least one plump bud, and preferably two. Replant, leaving 3ft/90cm between plants and rows, with the top of the buds no more than 1.5in/3cm below the level of the soil. Pick no stalks the first year.

Seed is sown in early spring in a frame, or outdoors in spring, in drills 0.5in/1.25cm deep. As soon as the seedlings can be handled, transplant them to a nursery bed, 6–8in/15–20cm apart. Move to their permanent bed the following spring or autumn.

Strawberry

RUNNERS WITH PLANTLETS

Fragaria x *ananassa*. Strawberries are dangerously easy to propagate — they do it themselves. This can be undesirable because they may also spread the virus diseases to which they are prone. It is possible to buy certified virus-free plants but that is no guarantee that they will stay completely healthy. Take precautions by starting a little bed just for propagation, well away from the bed for cropping. Renew it after 2, or at the most 3, years with fresh certified stock in another part of the garden.

Runners are propagated in summer to late summer, using bought-in certified virus-free runners. If planting an ordinary bed at the same time as the propagating bed, choose a few of the strongest-looking plants for raising replacement runners. Plant 15–18in/37.5–45cm apart. In the following year, remove all flowers from the plants in the propagating bed; their job is to produce runners, not fruit. The runners may root themselves, or can be pinned down with bent wires into the soil. In late summer they should be well rooted. Transplant them to fill gaps in the main bed, or start a new one. Everbearing strawberries (autumn fruiting) can also be propagated by runners. The best will

come from plants that have not been allowed to fruit. During the first winter they may look dead, but in early spring should begin to show signs of life. Plant out in suitable weather in their fruiting bed, 18–24in/45–60cm apart, according to the vigour of the variety.

In summer or late summer, pin down runners either into pots (for ease of transplanting) or into the soil.

Strawberry, Alpine

SEED; DIVISION

Fragaria vesca semperflorens, miniature strawberries, do not produce runners.

Seed is sown in late winter or early spring in a little warmth, or in a cold frame in spring. Sow in seed compost, but do not cover the seed. Treat as a half-hardy annual. Germination is unpredictable: the first seedlings may appear in 2 weeks; the more dilatory can take 2 months. Prick out into potting compost, 2in/5cm apart, as and when they are large enough to handle. Plant out in the garden in late spring, 12–18in/30–45cm apart, with 12in/30cm between rows. Expect fruit in late summer.

Division takes place after fruiting is over. Split up the clumps and replant. Some protection is advisable during winter.

VIRUS DANGER

The importance of propagating only from healthy plants of soft fruits cannot be over stressed.

The main threat comes from virus diseases, and it is often hard to diagnose them at an early stage. Even early diagnosis would not help for it is impossible to cure an affected plant. It must be burnt; not thrown on the compost heap.

The first defence against viruses is to buy healthy stock, preferably certified as being virus free. This is most important of all with strawberries, which are especially susceptible to virus disease.

The second defence is to wage war on pests which spread virus diseases. Aphids are particular foes of strawberries, and raspberry viruses are spread by aphids and eelworms. The blackcurrant suffers from a deadly reversion virus, largely spread by blackcurrant gall mite. Red currants and white currants are less affected by virus disease. In the United States blackcurrants are not grown because they are hosts to a fungus which damages forest pine trees.

The third line of defence, after destroying all affected plants, is to buy in new healthy plants and grow them in a different part of the garden. Then you will be able to start propagating again.

The Propagator's Calendar

A seasonal guide to
sowing and multiplying plants

WINTER

Bring dahlias and chrysanthemums into the warmth to start into growth. When young shoots are about 2in/5cm long they can be used as **basal cuttings**.

Sow, in heat, those half-hardy perennials which are to be grown as half-hardy annuals.

Sow onions, leeks, and early cauliflowers in a heated greenhouse.

There is little to do outside except to **plant** garlic, weather permitting — the sooner, the better.

LATE WINTER

Sow the bulk of half-hardy annuals in heat.

Alpine strawberries can be **sown** in heat this month or next.

Divide and replant clumps of snowdrops immediately after flowering. They are an exception to the general rule of waiting to divide bulbs until the leaves have died down.

EARLY SPRING

To raise some annuals for early flowering, start them off indoors. **Sow** the seeds in heat, a few weeks before they would be sown outdoors. Some annuals resent being transplanted, however (see individual plant entries).

In very mild areas, you can start **sowing** very hardy annuals outdoors from the middle of the month. But beware being over hasty, especially if the soil is cold and heavy. In such conditions seeds are more likely to rot than to germinate.

Divide herbaceous perennials as the soil begins to warm up.

Take **cuttings** from very young growth of pelargoniums, fuchsias, impatiens and other indoor plants.

This is also the time to **divide** many indoor plants as they are being repotted or potted-on.

Sowing begins of vegetables in the open or under cloches. For these early sowings, germination will be more rapid if cloches have been placed over the seed bed two weeks or so before sowing.

Sow aubergines and sweet peppers indoors.

SPRING

Now begins the main outdoor **sowing** time for hardy flowering annuals. No time should be lost when the ground is friable and the weather good.

While the sowing of indoor plant seed is less dependent upon the weather (assuming there is adequate artificial warmth) the longer hours of daylight in spring give seedlings a better start in life. They have a good chance to become well established before the autumn and winter months.

Divide herbaceous perennials, tubers and rhizomes.

Take **basal cuttings** from young tender growth of herbaceous perennials, rooting them indoors or in frames and cloches.

Transplant the **eye cuttings** of ornamental vines which have been in the propagator since early winter or winter. Gradually harden them off before moving to a cold frame.

Serpentine layering of some climbing plants can be carried out from now to late summer, whenever the shoots are long enough.

All hardy vegetables can be **sown**, including repeat sowings of those made earlier in the year. The routine of successional sowings of salad vegetables should be established and continued through the coming months.

Sow sweet corn indoors.

LATE SPRING

Plant out half-hardy annuals which have been raised indoors.

Finish **sowing** of annuals for summer flowering.

Sow biennials from now to late summer. In autumn, plant out in the garden to flower next year.

There is still time to make further **sowings** of vegetables. These may be successional crops, or first sowings outdoors of such tender vegetables as French beans. Many of the earlier sown vegetables will be ready for transplanting.

Take **softwood cuttings** of herbaceous perennials, from now to early summer.

EARLY SUMMER

Some annuals can still be **sown** to provide a succession of flowers into the autumn.

There are also more vegetables to be sown or transplanted. Delaying too long before sowing may not allow the plants time enough to mature.

Pin down strawberry **runners** for new plants.

SUMMER

From now onwards, even the most enthusiastic propagator can relax considerably. There is far less sowing to be done and new jobs will mainly involve taking **cuttings** in various ways. There will still be plenty to do in caring for all the new plants which have so far been propagated (not to mention everything else in the garden).

Sow cabbages to be cropped in spring next year.

Collect and plant stem **bulbils** of lilies. They will need to be kept in a cold frame for two years.

Semi-ripe cuttings, mainly of shrubby plants, can be taken between this month and early autumn.

Layer carnations and pinks.

Tip layer blackberries this month or next.

LATE SUMMER

Make final **sowings** of biennials.

Take **hardwood cuttings** of roses.

Plant out rooted **runners** of strawberries.

Take **cuttings** of rosemary, sage and mint.

EARLY AUTUMN

Sow sweet peas in frame for early flowers next year.

Plant out biennials and perennials raised from seed.

Plant spring flowering bulbs.

Take **cuttings** of gooseberries, blackcurrants, red and white currants, from now until autumn or late autumn.

AUTUMN

This is the main season for taking **hardwood cuttings** of deciduous flowering shrubs.

Plant out well rooted cuttings of soft fruit — canes and shrubs.

Prepare for next year by digging the ground where vegetables are to grow next spring.

LATE AUTUMN

Dig up and plant raspberry suckers.

Slack, dull days, the time to stay indoors and decide what to propagate when warmer weather arrives.

EARLY WINTER

Full dormancy has set in. Take **eye cuttings** from dormant stems of *Vitis cognitiae* and plant in heat, or wait until winter.

Meanwhile make your way critically through catalogues of seedsmen and nurserymen. Send off orders.

SEASONAL REMINDER

Throughout the book the months have been converted to seasons which are applicable to both Northern and Southern hemispheres.

Northern hemisphere		Southern hemisphere
January	Winter	June
February	Late winter	July
March	Early spring	August
April	Spring	September
May	Late spring	October
June	Early summer	November
July	Summer	December
August	Late summer	January
September	Early autumn	February
October	Autumn	March
November	Late autumn	April
December	Early winter	May

INDOOR PLANTS

Most people feel they have a much more intimate relationship with their indoor plants than with the plants in their garden. Plants outdoors tend to merge into a background of colour and greenery, while in bad weather they may seem only to exist as a rain-blurred image through the window. Indoors you maintain closer contact with your plants. You brush past them; you dust round them; you give them water. They are there when your eyes look up from a book or the TV. Each one in its pot is an individual, with its own place in the room. Even grouped together they preserve a greater individuality than the more anonymous members of a garden border or bed. So, by having plants around, you begin to understand them. People may claim that their houseplants do well because they love them, or talk to them, but it is simply because they notice what is happening to them. Knowledge painlessly absorbed by such observation helps not only to keep the plants alive and healthy, but vastly increases their chances of being successfully propagated by their owners.

Do heed a word of caution, however. Propagation of indoor plants should be undertaken with discreet enthusiasm. Otherwise you will gradually fill every room in the house with potted plants, and find your friends reluctant to accept any more from you.

HOW TO PROPAGATE

The same methods of propagation apply to indoor plants as for those outdoors, but there is a major difference. Many indoor species are not the hardier

Some popular houseplants are only too easy to propagate; others develop slowly. Those shown here, including *Impatiens* (busy lizzie), *Tradescantia* (rooting in water), *Streptocarpus* (right) and *Begonia rex* with its leaf cuttings, should recreate themselves without problems.

Softwood cuttings

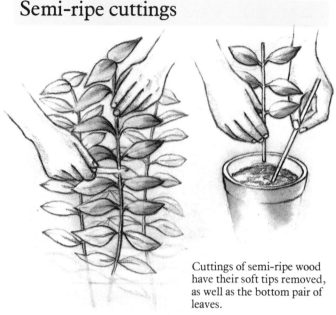

Take cuttings 3–4in/7.5–10cm long. Remove the lowest pair of leaves and plant several in a pot of compost.

Semi-ripe cuttings

Cuttings of semi-ripe wood have their soft tips removed, as well as the bottom pair of leaves.

plants of temperate climates, but tender exotics of tropical or sub-tropical origin. For them, warmth means being much warmer than is necessary for annuals, biennials and perennials in the open garden. So while methods of propagation may be the same, the required environments are not, and a heated propagator becomes indispensable for propagating the majority of these house and conservatory plants.

DIVISION

For many plants which grow in clumps, division is the simplest way to gain more plants, or to make a new start with younger growth from an ageing plant which has passed its best. Have clean pots ready so that the divided pieces can be replanted at once.

First, turn the plant out of its old pot by placing one hand over the surface of the compost, with fingers around or among the stems. Turn the pot upside down and tap the rim against the end of a table. A stubborn root ball can be released by running a knife round the inside of the pot. Gently disentangle the roots, trying not to damage too many root hairs. Tight clumps may have to be separated with a knife, but surgery causes more damage than careful pulling apart.

When repotting the divided portions, use fresh, moist compost of the same type as in the old pot. Spread a layer of compost in the new pot, up to the

level of the base of the roots. Hold the plant in one hand and, with the other, fill the pot evenly all round. Firm lightly, but allow some air to reach the roots. Divided plants are best kept out of the sun until they have settled down. Water when the compost begins to dry out.

STEM CUTTINGS

Stem cuttings taken from houseplants are usually of the softwood cutting and semi-ripe types. A softwood cutting is taken from the current year's growth during spring or perhaps early summer. Semi-ripe cuttings, which may root more readily, are taken between summer and autumn.

Softwood cuttings are taken with a sharp knife. Make a clean cut, as a jagged edge may start rot. Cut a shoot which has at least three nodes — the points at which the leaves join the stem — so that the cutting will probably be 3–4in/7.5–10cm long. Do not remove the tip, but trim the bottom to just below a node. Then take off the lowest pair of leaves, or more, so that a third to a half of the cutting is bare at the end.

To help induce rooting, use hormone rooting powder, which may also include a fungicide. Dip the bottom end of the cutting in water and then press it into the powder. Avoid pushing it in too far. The hormone will be absorbed mainly through the cut and

Plantlets and leaf cuttings

Leaf bud cutting

Choose a stem with a growth bud in the leaf axil. Trim the stem 1in/2.5cm above and 2in/5cm below the leaf.

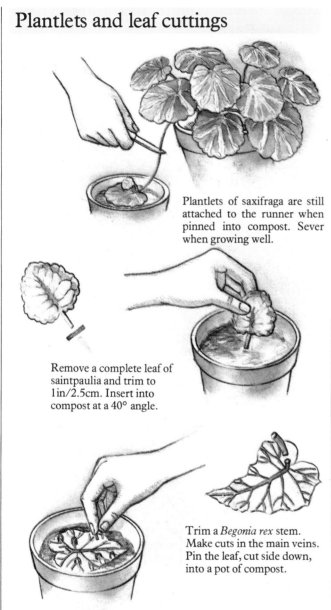

Plantlets of saxifraga are still attached to the runner when pinned into compost. Sever when growing well.

Remove a complete leaf of saintpaulia and trim to 1in/2.5cm. Insert into compost at a 40° angle.

Trim a *Begonia rex* stem. Make cuts in the main veins. Pin the leaf, cut side down, into a pot of compost.

any powder stuck to the stem itself may even harm later root growth.

bag, and open the ventilator in a propagator to prevent condensation forming. Keep cuttings out of the sun, but not in heavy shade. When they have obviously rooted — by showing signs of growth — gradually open the propagator more each day to accustom them to the lower temperature and drier air of the house.

Semi-ripe cuttings are treated in the same way as softwood cuttings, except that the soft tips are removed. Many will need less heat to root, and a higher proportion may actually do so as they are also less liable to wilt. Accept beforehand, however, that some cuttings will not root, but rot.

Stem cuttings of a different kind comprise short sections of stem, at least 1in/2.5cm thick. Pieces 3in/7.5cm long can be propagated if they have one or two nodes. They are usually planted horizontally in the compost and only half buried. Roots grow from the nodes in the compost; from those on top, shoots appear.

PLANTLETS AND LEAF CUTTINGS

Plantlets are good material for propagation. *Chlorophytum comosum, Saxifraga stolonifera* and *Tolmiea menziesii* conveniently grow their own plantlets, which can be layered and rooted. Other plants can be induced to grow plantlets from **leaf cuttings**. Whole leaves and stalks may be used, as with saintpaulias. Leaves

Cuttings should be planted quickly to avoid wilting. Insert them in pots or trays of compost, using a small dibber to make holes and so prevent damage caused by forcing cuttings in. In order to root, the cuttings need high humidity and continuous warmth, preferably from below, as provided by a base-heated propagator. Those which require only 60°F/16°C may root if kept inside a plastic bag, but for temperatures above that a propagator is more reliable. Regularly wipe off condensation from the inside of a and stalks are cut cleanly from the plant and the stalks trimmed to 1in/2.5cm. Dip the ends in hormone rooting powder, and plant to a depth of 0.5in/1.25cm in compost. The cuttings find better anchorage if inserted at an angle of 40° and appear to thrive better if arranged round the pot near the rim. The leaves should not touch each other or be in contact with the soil. In time, a plantlet grows at the base of the leaf and puts down roots. Transplant when well established. Leaves of *Begonia rex* and peperomias can also be propagated in this way.

Leaves with stalks removed can also be made to produce plantlets; this is the method often used to propagate *Begonia rex* and *Begonia masoniana*. Remove a leaf and cut off the stalk. Turn it underside up and cut into the main veins with a sharp knife. Turn the leaf over again and place it, cut side down, on the surface of a tray of moist compost. Pin it down with pieces of bent wire so that the leaf surface is in close contact with the compost. Place in a heated propagator and plantlets will appear from slashes in the veins. Separate from the leaves when large enough and move into individual pots of potting compost.

Other leaves will also produce plantlets from veins which run the length of each leaf. A sansevieria can be cut into 2in/5cm lengths, each capable of growing a plantlet in two months or more. This will only happen, however, if the pieces are inserted the same way up as they were growing on the plant.

Leaf-bud cuttings make use of leaves for yet another form of propagation. Each cutting consists of a leaf (a food source until new roots grow), a bud in the axil of the leaf (from which the new plant will appear) and a piece of stem (from which the roots will grow). *Ficus elastica* and *Ficus lyrata* can be propagated in this way; it may also be used for dracaenas.

LAYERING

Simple layering involves burying a stem behind its growing tip to induce the growth of roots. It is widely used outdoors, but few houseplants are suitable for this treatment.

Air layering, not greatly used outdoors, is a convenient method for houseplants. It is most often employed, however, not to produce more plants but to rejuvenate old ones which have grown leggy with long bare stems. *Ficus elastica* is often shown as an example, but codieaums, dieffenbachias and dracaenas are also suitable for this treatment.

Acalypha

SOFTWOOD CUTTINGS

Perennial shrubs, growing to 6ft/1.8m or more. Temperatures required: summer, 65–70°F/18–21°C; winter, not below 60°F/16°F. *Acalypha hispida*, Chenille Plant, has red, tassel-like flowers. *Acalypha wilkesiana* has brilliant coppery foliage.

Softwood cuttings are taken in spring. Because these acalyphas are so quick-growing, it is sensible to take cuttings every year to replace the less attractive second-year plants. Take stem tip cuttings about 4in/10cm long. Plant in 3in/7.5cm pots in peat potting compost with added sharp sand. Put in a propagator with a temperature of 75°F/24°C. When new growth shows, increase ventilation in the propagator. Keep the compost just moist. Move to 4in/10cm pots in potting compost when plants are about 10in/25cm tall.

Achimenes

DIVISION OF RHIZOMES; CUTTINGS; SEED; TUBERCLES

Perennial gesneriads. Profusely flowering plants, some species of which grow erect to 6in/15cm; some trail at 20in/50cm; some grow from small rhizomes which are dormant in winter. Temperatures

required: summer, 60–70°F/16–21°C; winter (when dormant), 50°F/10°C. *Achimenes longiflora*, Hot Water Plant, has trailing stems some 20in/50cm long, smothered in blue (or white) flowers from June to September.

Division of rhizomes takes place in early spring, when repotting, and is the easiest way to propagate this plant. Separate and break off some of the rhizomes, but do not cut. Plant each one in a small pot of gesneriad potting compost, 1in/2.5cm below the surface. Keep in 65°F/18°C and a new shoot will appear.
Softwood cuttings are taken in early summer. Use cuttings 4in/10cm long and plant in gesneriad potting compost. Keep in a propagator at 65°F/18°C, in bright indirect light.
Seed is sown in late winter to early spring in a tray of finely sieved gesneriad potting compost. The seed is minute and must not be covered as it needs light to germinate. Place in a propagator in a temperature of

RAISING PLANTS FROM SEED

Houseplants can, of course, be grown from seed, although this demands more heat than is required for vegetative propagation of tender plants. Hence the need for a heated propagator. Choose one heated from below (the seeds germinate better) and with a tall cover (to accommodate cuttings as well).

By providing adequate warmth in this way, the sowing season for indoor plants can be made to last from winter onwards.

Seed trays, in various sizes and 1.5–2in/3.5–5cm deep, are preferable to pots, which take up a lot of room in a propagator. Several plants can be grown in the same tray, if clearly labelled. Use loam or peat based composts, as recommended for individual plants. Fill the tray to the top with compost and press down with a flat piece of wood to level. Firm it a little. This should leave the compost about 0.25in/0.5cm from the top of the tray.

Sow seeds in mini-rows in the tray, 1in/2.5cm apart and shallow for small seeds to 2in/5cm apart and deeper for large ones. Form the rows by pressing pieces of cane across the compost and to the depth required.

After sowing thinly and evenly along the rows, cover seeds with a thin layer of sieved compost, unless the seeds are described as needing light for germination. Then water from below. Stand the tray in a bowl of shallow water and allow the water to soak through the compost. Then let the tray drain before putting it in the propagator.

Set the thermostat to the temperature recommended; many are in the range 65–70°F/18–21°C, but some will be higher. When numerous seedlings appear, lower the temperature, say by 5°F/3°C, and open the propagator ventilators a little. The seedlings now need good light, but not direct sunlight. Water when necessary from below.

Thin out overcrowded seedlings and when large enough to handle (after true leaves have grown), prick out in other containers with fresh compost.

Return the pricked out seedlings to the propagator and gradually begin to harden off, by admitting more air and lowering the temperature. When the seedlings are large enough to transplant, move into individual pots in the recommended potting composts. If there is room, return the pots to the propagator and gradually lower the temperature to room levels. Then bring the young plants out into their new world.

65–70°F/18–21°C. Germination takes 3–4 weeks, perhaps more, and during this time the compost must be kept moist. Some 8 weeks after the seedlings emerge it will be safe to transplant them into individual pots of potting compost.
Tubercles are planted in late spring. These small lumps are removed from the rhizomes and are less demanding to grow than seeds. Plant 1in/2.5cm deep in potting compost, allowing 4 tubercles to a 4in/10cm pot. Keep around 65°F/18°C and use slightly warm water for watering.

Gesneriad compost

Achimenes, a member of the gloxinia family, *Gesneriaceae*, requires a gesneriad compost. This is made in the proportions of equal parts of:
Sphagnum peat moss
Perlite
Vermiculite

Anthurium

DIVISION

Anthurium scherzerianum

Perennial flowering plant, growing up to 16in/40cm tall. Temperatures required:

summer, 65–70°F/18–21°C; winter, not below 60°F/16°C. *Anthurium scherzerianum*, Flamingo Flower, has dark green, leathery, lance-shaped leaves on 6in/15cm stems and, in spring and summer, produces a bright red spathe or bract surrounding an orange-red spadix, the flower spike. *Anthurium andreanum*, Painter's Palette, is larger, with heart-shaped leaves and pink to red spathes surrounding yellow spadices.

Division takes place in early spring. When repotting, break the root ball into 2 or 3 pieces, gently prising apart the roots and stems. Each piece must have plenty of roots to support the stems and leaves. Plant each piece in a pot of peat potting compost with added sphagnum moss. Keep at a steady 70°F/21°C, ideally in a propagator, and in a slightly shaded spot until the plant recovers from the shock and shows healthy new growth.

Aphelandra

SOFTWOOD CUTTINGS

Evergreen flowering shrub, reaching 18in/45cm. Temperatures required: summer, 65–70°F/18–21°C; winter, 60–65°F/16–18°C but tolerates temperatures down to 55°F/13°C. *Aphelandra squarrosa* 'Louisiae', Zebra Plant, has bright green, shiny, elliptic leaves with white veins. In summer, yellow flowers emerge from yellow overlapping bracts tipped with orange, arranged in a pyramid shape. 'Brockfeld' is a smaller variety, with veins picked out in cream, but its flowers and bracts are similar to those of 'Louisiae'.

Softwood cuttings are taken in late spring. Aphelandras are often past their best after a year, so it is wise to raise new plants annually. Take stem tip cuttings about 4in/10cm long, with at least 2 pairs of leaves. Remove the bottom pair and plant the cutting in peat potting compost. Water and place pots in a propagator. Keep at 70°F/21°C. Cuttings should root within 8 weeks.

Asparagus

DIVISION; SEED

Perennial evergreen, trailing to some 15in/37.5cm. Temperatures required: summer, 55–65°F/13–18°C, with higher temperatures tolerated; winter, 50–55°F/10–13°C. *Asparagus densiflorus*, Emerald Feather, has trailing stems with many small green phyllocades (flattened stems), prickly when mature. *Asparagus plumosus* 'Nana', Asparagus Fern, is compact, more feathery, and bright green. *Asparagus falcatus*, Sicklethorn, is a climber with sickle-shaped leaves which grows to 10ft/3m and more.

Division takes place in early spring. Remove plant from its pot and, with a sharp knife, divide the root ball into 2 or 3 sections, each with plenty of roots and foliage. Plant in loam potting compost. If top growth is poor, cut all stems down to compost level, remove the root ball and, as though slicing a cake, cut it into 4in/10cm sections. Plant each section in a pot of loam potting compost. New growth

will soon appear.

Seed is sown in spring. Prepare a tray with seed compost and scatter seed evenly. Sprinkle with a layer of compost only to the depth of the seed. Cover tray with polythene or place in a propagator at 60–65°F/16–18°C. Keep the compost moist. Germination should take place in about 4 weeks. Thin out seedlings to 1in/2.5cm when large enough to handle. Transplant to individual pots of loam potting compost when established.

Aspidistra

DIVISION

Perennial foliage plant growing up to 2ft/60cm tall. Temperatures required: summer, 60°F/16°C; winter, 50°F/10°C, but tolerates up to 60°F/16°C. *Aspidistra elatior*, Cast Iron Plant, has dark green, leathery, lance-shaped leaves on short stems. 'Variegata' has white to cream stripes on leaves.

Division of *Aspidistra eliator*

Division takes place in spring. When repotting, gently break away compost to reveal the rhizome from which leaves

emerge. Cut away pieces of rhizome with 4 or more leaves and plenty of roots. Plant several pieces in a pot of loam potting compost. If pieces of rhizome without leaves should break off, these can still be used for propagation as long as they have buds. Plant with the buds pointing upwards and outwards. Eventually they will put out leaves.

Aucuba

SEMI-RIPE CUTTINGS

Evergreen shrub growing up to 6ft/1.8m high, but most indoor varieties are no more than 3ft/90cm. Temperatures required: summer, 55–60°F/13–16°C; winter, 45°F/7°C but slightly higher temperatures will be tolerated. *Aucuba japonica* 'Variegata', Spotted Laurel, has shiny serrated leaves, blotched with yellow.

Semi-ripe cuttings are taken in late summer. Take cuttings about 6in/15cm long with at least 2 pairs of leaves and a healthy growing point. Remove the lowest pair of leaves. Place cutting in a mixture of peat and coarse sand, water and cover with a plastic bag or place in a propagator. Keep between 65–70°F/18–21°C in a slightly shaded spot. When new growth appears, remove the plastic bag or take cuttings from the propagator. When the plant is about 12in/30cm high, transfer to a pot of loam potting compost.

Begonia

SEED; SOFTWOOD CUTTINGS; DIVISION OF TUBERS; LEAF CUTTINGS

Perennial foliage and flowering plants which grow up to 18in/45cm high, depending on the species. Temperatures required: summer, 65–70°F/18–21°C; winter, 55–60°F/13–16°C. *Begonia rex*, Painted Leaf Begonia, has heart-shaped green leaves with multi-coloured markings. *Begonia masoniana*, Iron Cross Begonia, has green puckered leaves with a brown-red cross. *Begonia tuberhybrida*, Tuberous Begonia (dormant in winter) has long, heart-shaped leaves and produces single or double flowers in white, yellow, orange and crimson, from early summer to early autumn. *Begonia semperflorens* has heart-shaped green, red or purple leaves with white, pink or red flowers, from early summer to early autumn. 'Gloire de Lorraine' hybrids have pink to red flowers, from late spring to early autumn. 'Schwabenland' hybrids have larger leaves and may produce orange, red or yellow flowers off and on throughout the year.

Begonia rex

Seed is sown in early spring for *Begonia rex*, *Begonia semperflorens* and *Begonia tuberhybrida*. Fill a tray with seed potting compost and scatter the fine seed evenly over the surface. Do not cover. Water the compost gently and put the tray in a propagator. Keep between 70–75°F/21–24°C. *Begonia rex* and *Begonia semperflorens* take about 4 weeks to germinate; *Begonia tuberhybrida* may take up to 8 weeks. When large enough to handle, thin out seedlings to 1in/2.5cm. Later transplant to pots of peat potting compost.

Softwood cuttings are taken from spring through summer for *Begonia semperflorens*, *Begonia tuberhybrida*, 'Gloire de Lorraine' and 'Schwabenland.' Cuttings should be about 4in/10cm long, with at least 2 pairs of leaves. Remove the lower pair and plant in a mixture of peat and sand. Place pot in a propagator and keep at 70°F/21°C until new growth appears, usually 6 weeks later. Some 4 months later transplant into a pot of peat potting compost.

Division of tubers to raise *Begonia tuberhybrida* takes place in spring. In winter, dormant tubers are stored dry at 45°F/7°C. In spring, cut the tuber into pieces, ensuring that each piece has a growing shoot. Plant in pots of peat potting compost.

Leaf cuttings (with stalks) To raise *Begonia rex* and *Begonia masoniana*, in early summer remove a leaf with about 2in/5cm of stalk attached. Insert the stalk at an angle of 45° in a pot of half peat compost and half sand. Place in a propagator and keep at 70°F/21°C. After some 6 weeks, small plantlets should appear around the base of the stalks. Let these become well established for a further 4–6 weeks before planting in pots of potting compost.

Leaf cutting of *Begonia rex*

Leaf cuttings (without stalks) Take a complete leaf and place on a flat surface, *underside upwards*. Make 2 or 3 cuts with a sharp knife into the central vein and the larger veins joining it. Be careful not to cut right through. Partly fill a tray with damp sand and place the leaf, *face side upwards*, on the surface. Secure on to the

sand with loops of wire and place in a propagator at 70°F/21°C. After about 6 weeks, plantlets should appear from the cuts in the veins. When each plantlet has at least 2 leaves of 1in/2.5cm, sever from the parent leaf and plant in peat potting compost.

Brassaia

SEED; AIR LAYERING

Perennial shrub reaching 6ft/1.8m. Temperatures required: summer, 60–65°F/16–18°C; winter, 60°F/16°C, and not below 55°F/13°C. *Brassaia actinophylla* (syn. *Schefflera actinophylla*), Queensland Umbrella Tree, has a central stem with stalks bearing 8 or more bright green oval leaflets in umbrella formation. *Brassaia digitata* grows more delicate and graceful leaflets.

Seed is sown in early summer. Fill a tray with loam seed compost and scatter seed evenly over the surface. Lightly cover with compost to the depth of the seed. Put in a propagator and keep at 70°–75°F/21–24°C in a shaded spot. Thin out seedlings to 1in/2.5cm and when established transfer to pots of loam potting compost.

Air layering gives new life to large brassaias which have lost their lower leaves and are past their best. Make an upward slanting cut with a sharp knife a little way into the main stem, about 2in/5cm below the lowest leaf. Dust the cut with hormone rooting powder and wedge it open with a matchstick. Cover with damp sphagnum moss, securing this in place with string. Wrap the moss in clear plastic, firmly tied round the stem, top and bottom, to keep in the moisture. Continue watering and feeding the plant as usual. After about 8 weeks, roots should appear growing through the moss. Remove the plastic, cut the stem just below the moss and plant roots and moss in loam potting compost. The bare lower stem will also put out new growth.

Caladium

DIVISION OF TUBER

Foliage plant which grows to 2ft/60cm. Temperatures required: summer, 70°F/21°C; winter, 60°F/16°C (tubers are dormant, stored dry in compost). *Caladium hortulanum* hybrids, Angel Wings, have paper thin arrowhead leaves of white, green, pink or red, with veins of green and red.

Division of tuber takes place in early spring. Turn the dry soil ball out of the pot and gently break away compost. Detach offset tubers, making sure that each has a growing point and some roots. Plant in 3in/7.5cm pots of peat potting compost to a depth which equals that of the tuber. Put in a propagator and keep at 70°F/21°C. When tubers start to grow, take pots from the propagator, but keep at 70°F/21°C.

Calathea

DIVISION

Perennial foliage plant which grows up to 2ft/60cm high. Temperatures required: summer, 60–70°F/16–21°C; winter, the same, and never below 60°F/16°C. *Calathea makoyana*, Peacock Plant, has broad oval leaves, green with silvery streaks embossed with a pattern of 'leaflets'; the undersides are purple. *Calathea insignis* (syn. *Calathea lancifolia*), Rattlesnake Plant, has erect, lance-shaped, light green leaves with olive markings. *Calathea zebrina* has leaves of emerald green with lighter green mid-rib and veins.

Division takes place in late spring. When repotting, large plants can be divided. Remove plant from the pot and gently prise apart the root ball into 2 or 3 sections, ensuring that each piece has enough roots to support the foliage. (If plants are divided into small pieces it will be many months before they achieve a good shape.) Plant in peat potting compost and keep in a shaded spot to let them recover.

Calceolaria

SEED

Annual flowering plant which grows up to 14in/35cm, but many hybrids are 6–9in/15–22.5cm in height. Temperatures required: summer, 60–70°F/16–21°C; winter, 45–50°F/7–10°C, for overwintering of plants raised from seed sown in autumn. *Calceolaria multiflora*, Slipperwort, has oval, mid-green leaves and blousy, pouch-like, multi-coloured pink, orange, red and yellow flowers from spring to early autumn.

Seed is sown in autumn for early flowering and for later flowering in early spring to spring. Fill a tray with seed compost and scatter the fine seed over the surface. Do not cover, but gently water seeds into the compost. Put the tray in a propagator and keep a temperature of 70–75°F/21–24°C for germination, which takes about 3 weeks. Thin out the seedlings to 1in/2.5cm when they can be handled. Gradually harden off and, when well established, transplant into pots of peat potting compost. Plants should flower 4 months or more after germination.

Callisia

SOFTWOOD CUTTINGS; BASAL CUTTINGS

Perennial foliage plant which trails to 2ft/60cm. Temperatures required: summer, 60–65°F/16–18°C; winter, 55°F/13°C. *Callisia elegans*, Striped Inch Plant, has dark green oval leaves with white stripes on the upper surface and purple undersides.

Softwood cuttings are taken from spring to early summer. Remove the lower leaves from 4in/10cm cuttings, exposing about 1.5in/3.75cm of stem. Insert several in a pot of peat or loam potting compost and keep warm in a slightly shaded pot. They will normally root in about 3 weeks. Cuttings will also root in water and can be moved into compost, but there is little point in choosing this method, as there is an inevitable setback with the move. Propagate callisias every year as they are usually past their best after 2 years.

Basal cuttings are the new shoots which appear from the base of the plant. From spring to early summer take 3in/7.5cm cuttings with a 'heel', close to the surface of the compost. Remove the bottom 2 leaves and then treat in the same way as for softwood cuttings.

Chamaedora

SEED; DIVISION (SUCKERS)

Palm growing 4ft/1.2m tall with fronds 2ft/60cm long. Temperatures required: summer, 65–70°F/18–21°C; winter, minimum of 55°F/13°C. *Chamaedora elegans*, Parlour Palm, has arching, greenish-yellow stems with pairs of narrow leaflets. Yellow, mimosa-like flowers appear in summer.

Division of *Chamaedora elegans*

Seed is sown in spring, but growth is a long process. Soak the hard-coated seed for at least 24 hours before sowing. Alternatively, scratch the surface of the seed, but do not dig deeply into it. Sow in a 3in/7.5cm pot of peat seed compost. Cover with compost to the seed's own depth, and place in a propagator at 75–80°F/24–27°C. Germination may take up to 6 months. Let the delicate seedlings grow to about 3in/7.5cm before transplanting to individual pots, using two thirds peat potting compost to one third sand. After they recover from transplanting, move from the propagator and gradually acclimatise to lower room temperatures.

Division (suckers) When repotting in early spring, gently break away the old compost to see if any suckers (shoots) have developed from the main root. Break away suckers, at least 8in/20cm long, with plenty of roots to support them. Plant in two thirds peat compost with one part sand. Keep in a warm and slightly shaded spot until established.

Chlorophytum

DIVISION; LAYERING PLANTLETS

Perennial foliage plant with arching leaves up to 18in/45cm and runners to 4ft/1.2m. Temperatures required: summer, 55–65°F/13–18°C, and higher; winter, much the same and not below 45°F/7°C. *Chlorophytum comosum* 'Vittatum', Spider Plant, has clumps of thin, arching leaves with a cream streak. Runners bearing plantlets have scaly bumps at the base, from which roots will grow to produce replicas of the parent plant.

Division takes place in early spring. When potting-on, divide overcrowded clumps. Gently break away compost from the fleshy roots and split into 2 or 3 sections. Plant in pots of loam potting compost.

Layering plantlets takes place from spring to early summer. Pin down plantlets, still attached to the runners, in pots of loam potting compost. Just cover the base of the plantlet to induce roots to grow. After about 6 weeks the plantlets should be growing well and the runners may be cut to detach them from the parent plant.

Cissus

SOFTWOOD CUTTINGS

Perennial evergreen climbing and trailing plant which grows to 10ft/3m. Temperatures required: summer, 55–65°F/13–18°C; winter, 55°F/13°C

and no lower than 50°F/10°C. *Cissus antarctica*, Kangaroo Vine, has dark green, oval, pointed leaves with serrated edges. *Cissus striata* has dark green leaves of 5 leaflets with pink undersides, on red stems. *Cissus rhombifolia* has rhomboid-shaped dark green leaves.

Softwood cuttings are taken from spring to early summer. Take stem tip cuttings about 4in/10cm long. Remove the lowest pair of leaves and plant several cuttings in a 3in/7.5cm pot of equal parts of loam and sand. Cover with a plastic bag or put in a propagator. Keep at 65°F/18°C in a slightly shaded spot. Within 8 weeks or so the cuttings should root and new growth will show. Remove pot from the bag or propagator. Pot-on when roots appear to fill pot, using a loam potting compost or a mixture of loam, peat, leaf mould and sand.

Clivia

DIVISION (OFFSETS)

Perennial flowering plant. Its leaves are up to 2ft/60cm long; spread to 3ft/90cm.

Division of *Clivia miniata*

Temperatures required: summer, 65°F/18°C; winter, 50–55°F/10–13°C. *Clivia miniata*, Kaffir Lily, has strap-shaped, leathery, dark green leaves in arching pairs. A cluster of trumpet-shaped orange flowers on a long stem appears in early spring or spring.

Division takes place after flowering with a clivia that has produced a decent-sized offset, or offsets. Remove the plant from its pot and select offsets with at least 3 leaves about 9in/22.5cm long. Carefully disentangle the fleshy roots and cut the offset, or offsets, from the main plant. Each offset must have plenty of roots or growth will be both slow and poor. Plant each offset in a pot of loam potting compost. It should flower after 3 years, but could take longer.

Codiaeum

SOFTWOOD CUTTINGS; AIR LAYERING

Perennial shrub which grows to 3ft/90cm. Temperatures required: summer, 60–70°F/16–21°C; winter, similar to those of summer but not below 55°F/13°C. *Codiaeum variegatum pictum*, Croton, has a variety of leaf shapes: broad, oblong, fingered or lobed; in combinations of yellow, green, pink, brown, orange and black.

Softwood cuttings are taken in late spring from stem tip cuttings about 4in/10cm long, with at least 2 pairs of leaves. Remove the bottom pair and plant in a 3in/7.5cm pot of half peat and half sand. Place in a propagator and keep at 75°F/24°C. After about 6 weeks, new growth should be seen and the pot can be moved from the propagator. When pot is full of roots, pot-on, using a peat potting compost.

Air layering may be undertaken in late spring. It is an ideal method for codiaeums which have lost their lower leaves. Make an upward slanting cut into the main stem, about 2in/5cm below the lowest leaf. Do not cut too far into the stem. Wedge open the cut with a matchstick, and dust with hormone rooting powder. Then continue as instructions for air layering a Brassaia, see page 113.

Coleus

SOFTWOOD CUTTINGS; SEED

Perennial foliage plant, often treated as an annual, growing to 18in/45cm high. Temperatures required: summer, 60°F/16°C; winter, 55°F/13°C (for overwintered plants). *Coleus blumei*, Flame Nettle, has heart-shaped leaves with serrated edges. Leaves may be brown, green, yellow, red or orange, or combinations of these.

Softwood cuttings are taken in late summer from stem tip cuttings about 3in/7.5cm long. Remove the lowest pair of leaves and plant in loam or peat potting compost. Keep in a warm, slightly shaded spot out of any direct sun. The cuttings should root in about 3 weeks.

Seed is sown in early spring. Prepare a tray with loam seed compost and scatter seeds evenly over the surface. Do not cover, but gently water seeds into the compost. Place in a propagator at 65–75°F/18–24°C. Germination takes about 3 weeks. When seedlings can be easily handled, thin out to 1in/2.5cm. When they reach 2in/5cm, transplant to pots of loam potting compost.

Columnea

SOFTWOOD CUTTINGS; SEED

Perennial flowering plant which trails to 3ft/90cm. Temperatures required: summer, 65°F/18°C; winter, 60°F/16°C. *Columnea microphylla*, Goldfish Vine, has small, almost round, dark green leaves covered with reddish hairs. Red-tipped flowers with orange

throats appear in spring. *Columnea banksii* has fleshier leaves and scarlet flowers streaked with yellow. *Columnea gloriosa* has leaves with purplish hairs and scarlet flowers.

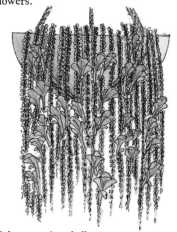

Columnea microphylla

Softwood cuttings are taken when flowering has finished. Take 4in/10cm stem tip cuttings and remove the lowest pair of leaves. Insert cuttings in a mixture of peat potting compost and sand. Place in a propagator at 70°F/21°C. In 3 weeks roots should develop. Remove from the propagator. When cuttings are well established, transfer several to a hanging container, planting in peat potting compost.

Seed is sown in early spring. Prepare a tray with peat seed compost and scatter seed evenly over the surface. Do not cover, but gently water seeds in. Put in a propagator at 75°F/24°C. Germination may take up to 2–4 months. Thin seedlings to 1in/2.5cm and when growing well, transplant several to a hanging container, using peat potting compost.

Cordyline

CUTTINGS (STEM SECTIONS); SEED

Evergreen shrub reaching 2ft/60cm tall. Temperatures required: summer, 70°F/21°C; winter, minimum 60°F/16°C. *Cordyline terminalis* 'Lord Robertson' has green leaves marked with red and crimson. The leaves of 'Prince Albert' variety have red markings.

Cuttings are taken in spring from a

cordyline which has become leggy, with a bare lower stem and a few leaves at the top. Cut the stem into 3in/7.5cm pieces, ensuring that each piece has several nodes. Fill a tray with a mixture of damp peat and sand and press the sections on to the surface. A node in contact with the compost will produce roots, while a node exposed to the air will grow a shoot. Put the tray in a propagator at 75°F/24°C and shoots should appear within 8 weeks. Harden off the cuttings over a period of 2 weeks to accustom them to lower room temperatures. Then pot into peat potting compost.

Seed is sown in spring. Fill a pot or tray with a mixture of half peat and half sand. Soak seeds in warm water for 3 hours before sowing 1in/2.5cm deep. Place in a propagator at 75–80°F/24–27°C. Germination takes 4–10 weeks. When seedlings are about 3in/7.5cm high, transplant to pots of peat potting compost.

Cutting of *Cordyline*

Cyclamen

SEED; DIVISION OF CORMS

Corm produces a plant which is 12in/30cm tall. Temperatures required: summer and winter, 45–55°F/7–13°C. Corms become dormant after flowering. *Cyclamen persicum* has heart-shaped light or dark green leaves, marbled with silver, bearing red, pink or white shuttlecock-shaped flowers in winter months.

Seed is sown in early spring. Soak seeds in hot water, leaving them to stand in it for a day or two before sowing. Fill a pot or tray with peat seed compost and sow seeds to their own depth. Cover with a plastic bag or put in a propagator at no more than 60°F/16°C. Germination takes place usually within 8 weeks. When seedlings are about 3in/7.5cm high, transplant to pots of peat potting compost.

Division of corms takes place in late summer. Before starting a dormant corm into growth, remove it from the pot of compost. Large corms can be divided in two, using a sharp knife. Each piece must have growing points and adequate roots. Plant in peat potting compost so that the corm just shows above the surface.

Cyperus

DIVISION; LEAF ROSETTES AS CUTTINGS; SEED

Perennial evergreen plant reaching 4ft/1.2m tall. Temperatures required: summer, 55–65°F/13–18°C; winter, 50–55°F/10–13°C. *Cyperus alternifolius*, Umbrella Plant, has tall stems carrying rosettes of green arching bracts. Brownish green flowers appear from the middle of the rosette in summer. 'Variegatus' has leaves striped with white and 'Gracilis' is a dwarf form.

Division takes place in spring. When

potting-on, break away old compost from the root ball. If faced with a mass of roots, you may have to cut a little way into the ball to make it easier to split. Very carefully pull the roots apart and divide the plant into 2 or 3 pieces. Pot each piece in loam potting compost.

Leaf rosettes as cuttings are taken in summer. Select some rosettes which have flowered and cut them off with 1in/2.5cm of the stem. Trim the bracts to half their length and root them either in shallow water or on a tray of sand which is kept constantly damp. Turn the rosette upside down, with the stem in the air, and place in the water or on the sand. Keep at 65–70°F/18–21°C in a slightly shaded spot. When roots develop, transfer to a pot of loam potting compost. A new plant will start to shoot from the middle of the rosette.

Seed is sown in spring. Fill a pot or tray with loam seed compost. Scatter seeds over the surface and cover them to their depth. Put in a plastic bag or propagator and keep at 65–70°F/ 18–21°C. Thin seedlings to 1in/2.5cm when they can be easily handled. When they reach 3in/7.5cm, transplant to loam potting compost.

Dieffenbachia

STEM TIP CUTTINGS; STEM SECTIONS; AIR LAYERING

Perennial foliage plant which grows 2–5ft/60cm–1.5m tall. Temperatures required: summer, 65°F/18°C; winter, minimum 60°F/18°C. *Dieffenbachia picta*, Dumb Cane, grows to 2ft/60cm and has lance-shaped leaves with white or cream markings; the larger *Dieffenbachia amoena* reaches 5ft/1.5m, with less pronounced markings.

Stem tip cuttings are taken in spring or late spring. Cut the stem just below a node and take cuttings about 6in/15cm long. Remove the lowest 2 leaves and plant in a pot of half loam potting compost and half sand. Place in a propagator at 70°F/21°C. It should root in about 6 weeks. Remove from the propagator and pot-on into loam potting compost when roots fill the old pot.

Stem sections are taken from bare, leggy

stems. Cut these off at compost level and divide into sections 3in/7.5cm long, each with several nodes. Place horizontally on the surface of a tray filled with half loam potting compost and half sand, well moistened. Place in a propagator at 70°F/21°C. Growth will start from the nodes. Transplant into loam potting compost.

Air layering of *Dieffenbachia picta*

Air layering is used for the much thicker stems of *Dieffenbachia amoena*. When these grow bare, make an upward slanting cut into the stem, 3in/7.5cm below the lowest remaining leaf. Avoid cutting too far into the stem. Wedge the cut open with a matchstick and dust with hormone rooting powder. Pack moist sphagnum moss in a ball round the cut and secure with string. Cover the moss with clear plastic, sealing it top and bottom with sticky tape to keep in the moisture. Water and feed the plant as usual. After 8–10 weeks the moss ball should be filled with roots. Cut the stem just below the moss ball and plant it, complete with the moss, in a pot of loam potting compost. New growth may also appear from the bare parent stem after a time.

Dizygotheca

SEED

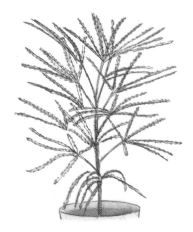

Perennial shrub which grows to 5ft/1.5cm. Temperatures required: summer, 65°F/18°C; winter, minimum 60°F/16°C. *Dizygotheca elegantissima*, False Aralia, has dark green leaves comprising several narrow glossy leaflets with serrated edges.

Seed is sown in spring. Prepare a pot or tray with peat seed compost. Scatter seeds evenly over the surface but do not cover; just gently water the seeds in. Put in a propagator at 70–75°F/21–24°C. Germination takes 3–4 weeks. When seedlings can be easily handled, thin to 1in/2.5cm. Transplant seedlings of 3in/7.5cm to individual pots of peat potting compost.

Dracaena

STEM TIP CUTTINGS; STEM SECTIONS; SEED

Evergreen shrub growing 18in/45cm-5ft/1.5m tall. Temperatures required: summer, 65-75°F/18-24°C; winter, 65°F/18°C, and no lower than 60°F/16°C. *Dracaena marginata* 'Tricolor', Dragon Tree, has rosettes of narrow, arching leaves striped green, cream and pink. *Dracaena deremensis* 'Bausei' has sword-shaped green leaves with a white stripe down the middle. 'Warneckii' has several white stripes on each leaf. *Dracaena fragrans* 'Massangeana' has a rosette of sword-shaped leaves striped with yellow. *Dracaena godseffiana*, the smallest, is no more than 18in/45cm, with oval, glossy green leaves spotted yellow.

Stem tip cuttings are taken in early summer. Remove the lowest pair of leaves from a 4–6in/10–15cm stem tip cutting and plant in a pot of half loam compost and half sand. Place in a propagator at 75°F/24°C. Within about 6 weeks new growth should be visible, showing that the cutting has rooted. Over a period of 2 weeks acclimatise the plant to lower room temperatures. When roots fill the pot, pot-on, using loam potting compost. This method is the only way to propagate *Dracaena godseffiana*.

Stem sections are taken from bare leggy stems. Cut these into sections about 3in/7.5cm long, ensuring that each piece has 2 or more nodes. (The top part of the stem which still has leaves can make a stem tip cutting, 4in/10cm or more long, as above.) To plant the stem sections, prepare a tray of half loam compost and half sand, well moistened, and place the sections horizontally on the surface. After 6 weeks or so growth will appear from the nodes, producing roots and shoots. When

Stem sections of *Dracaena marginata*

well established, transplant to pots of loam potting compost.

Seed is sown in early spring. Soak seed in warm water for 24 hours before sowing, or scratch the hard surface with a sharp knife, taking care not to damage the inner tissue. Plant in a pot of half loam compost and half sand, and place in a propagator at 75–85°F/24–30°C. Germination often takes up to 6 months. When seedlings are large enough to handle, transplant to individual pots. When young plants are about 6in/15cm tall, acclimatise over 2 weeks to lower room temperatures, then remove from the propagator. When roots fill the pots, pot-on, using loam potting compost.

Euphorbia

SOFTWOOD CUTTINGS

Perennial shrub which grows to 18in/45cm. Temperatures required: summer, 65°F/18°C; winter, similar to summer's but temperatures down to 55°F/13°C tolerated. *Euphorbia pulcherrima*, Poinsettia, has bright green, lobed leaves. Red, pink or white bracts surrounding small yellow flowers appear from early winter to late spring.

Softwood cuttings are taken in late spring or early summer. From cuttings about 4in/10cm long, remove the bottom pair of leaves and plant in a mixture of half peat potting compost and half sand. Place in a propagator at 70°F/21°C. The compost should never be very wet or the cuttings will rot. In about 8 weeks they should root. When roots are plentiful, move into pots of peat potting compost, disturbing the roots as little as possible.

xFatshedera

STEM CUTTINGS

Evergreen shrub growing to 4ft/1.2m. Temperatures required: summer, 55–65°F/13–18°C (the lower end of the range preferable); winter, 45–50°F/7–10°C. x*Fatshedera lizei*, Ivy Tree, has dark green, glossy lobed leaves. 'Variegata' has white to cream edged leaves.

Stem cuttings are taken in early summer

from the newly produced stems thrown up from the plant base. Use these rather than cuttings taken from the main stem; removing them spoils the appearance of the plant and the size of the leaves makes them difficult to cope with. Cuttings should be about 4in/10cm long. Remove the 2 bottom leaves and insert each cutting in a pot of half loam potting compost and half sand. Place in a plastic bag, supported by a wire hoop, and keep at 60–65°F/16–18°C in a slightly shaded spot. A heated propagator is not necessary. New growth should appear after about 6 weeks, showing that the cutting has rooted. Remove the plastic bag. When the pot is filled with roots, pot-on, using loam potting compost.

Fatsia

STEM CUTTINGS; SEED

Perennial shrub which grows to 5ft/1.5m, with a similar spread. Temperatures required: summer, 55–65°F/13–18°C (but the nearer to 55°F/13°C the better); winter, 45–50°F/7–10°C. Fatsia japonica, Japanese Fatsia, is a parent of xFatshedera, with leaves which are similar, but larger and more deeply incised. Fatsia japonica 'Variegata' has white to cream edged leaves and random blotches of contrasting colour.

Stem cuttings are taken and raised in exactly the same way as those of xFatshedera lizei above.
Seed is sown in spring. Prepare a pot or tray with a mixture of loam seed compost and sand. Sow evenly over the surface and cover with compost 1in/2.5cm deep. Put in a plastic bag and keep at 60–65°F/-16–18°C, in a good light, but out of direct sun. When seedlings can be easily handled thin, if necessary, to 1in/2.5cm. Transplant to pots of loam potting compost when 3in/7.5cm high.

Ficus

SOFTWOOD CUTTINGS; LEAF BUD CUTTINGS; AIR LAYERING; SEED

Perennial shrub which grows 3–6ft/90–180cm tall. Trailing varieties grow to 4ft/1.2m. Temperatures required: summer, 65–75°F/18–24°C; winter,

60°F/16°C. Ficus benjamina, Weeping Fig, has oval, pointed, dark green glossy leaves on arching stems. Ficus diversifolia (syn. Ficus deltoidea), Mistletoe Fig, has pear-shaped, leathery leaves and yellow berries. Two varieties of Ficus elastica — 'Decora' and 'Robusta' — have glossy, leathery, elliptical dark green leaves. 'Tricolor' has pink and cream markings, and 'Shrijveriana', pale green to yellow patches. Ficus lyrata (syn. Ficus pandurata), Fiddle Leaf Fig, has large, tough, fiddle-shaped leaves. Ficus pumila, Creeping Fig, is a trailer with stems bearing many small, heart-shaped leaves.

Leaf bud cutting of Ficus benjamina

Softwood cuttings of Ficus pumila, Ficus diversifolia and Ficus benjamina are taken in late spring. Take pieces about 4in/10cm long, making a cut just below a node. Remove the bottom pair of leaves and insert in a mixture of half peat potting compost and half sand. Place in a propagator at 75°F/24°C. When new growth indicates that cuttings have rooted, gradually acclimatise over 2 weeks to lower room temperatures. After about 5 months, roots will probably fill the pots.

Pot-on the cuttings, using peat potting compost.

Leaf bud cuttings of Ficus elastica and Ficus lyrata are taken in late spring but are not always successful. The method is worth a try, however, with a plant which has a bare stem and only a few leaves. The leaves chosen must have a growth bud where the leaf stalk joins the stem. Cut the stem about 1in/2.5cm above the leaf and 2in/5cm below it. Dip the bottom of the stem in hormone rooting powder and insert in a pot of half peat potting compost and half sand. See that the base of the leaf stalk is level with the surface of the compost. Support the leaf with a small cane to prevent it from toppling over and pulling out the piece of stem from the compost. Largish leaves can be curled into a cylinder and held in place with an elastic band. Place in a propagator at 75–80°F/24–27°C. The bud at the base of the leaf should start to grow after about 10 weeks. When the cutting is well established, move it into a pot of loam potting compost.

Air layering of Ficus elastica and Ficus lyrata takes place in late spring. This is another method for use only when a stem is bare and way past its best. Make an upwards cut a little way into the stem, about 2in/5cm below the lowest leaf. Dust the cut with hormone rooting powder and wedge it open with a matchstick. Pack moist sphagnum moss around the cut, in the shape of a ball, and secure in place with string. Cover with clear plastic (cling film is ideal) and bind the top and bottom with sticky tape to keep in moisture. Water and feed the plant as usual. In 8–10 weeks the moss ball should be full of roots. Remove the plastic covering and cut the stem just below the moss ball. Plant the rooted stem, complete with moss, in peat potting compost.

Seed is sown in spring. Prepare a pot or tray with peat seed compost. Scatter seeds evenly over the surface and gently water into the compost. Do not cover with compost because they need light to germinate. Place in a propagator at 70–80°F/21–27°F. Germination takes 6–12 weeks. Thin out seedlings to 1in/2.5cm when they can be easily handled. Transplant to pots of peat potting compost when about 3in/7.5cm tall.

Indoor Plants

Fittonia

SOFTWOOD CUTTINGS; LAYERING; DIVISION

Fittonia verschaffeltii

Perennial creeper which spreads to 12in/30cm and more. Temperatures required: summer, 65–70°F/18–21°C; winter, minimum 60°F/16°C. *Fittonia verschaffeltii*, Red Nettle Leaf, has oval, dark green leaves and carmine veins. *Fittonia argyroneura*, Silver Net Leaf, has silver to white veins.

Softwood cuttings are taken in spring. Use stem tip cuttings, about 3in/7.5cm long, with 4 or more pairs of leaves. Remove the bottom pair and insert in a pot of half peat potting compost and half sand. Place in a propagator at 75°F/24°C. When cuttings have rooted and are well established, remove from propagator. Pot-on when the container is filled with roots, using a mixture of two thirds peat potting compost and one third sand.
Layering takes place in spring. Pin down the growing tips of the creeping stems, while still attached to the parent plant, in pots of half peat potting compost and half sand. Within 6 weeks they should have rooted and can be severed from the parent. When growing well, transplant to pots of two thirds peat potting compost and one third sand, planting several to each pot to make a good show.
Division takes place in spring, when potting-on. The method is possible for large plants, but raising new young plants from cuttings is preferable. To divide, gently break away old compost from the roots. Split the root ball into 2 pieces and plant each in a pot of two thirds peat potting compost and one third sand. Keep warm in a slightly shaded spot until they recover from the shock.

Fuchsia

SOFTWOOD CUTTINGS; SEMI-RIPE CUTTINGS; SEED

Deciduous flowering shrub, which grows 12–18in/30–45cm tall. Temperatures required: summer, 55–65°F/13–18°C; winter, 50–55°F/10–13°C, but not below 45°F/7°C. *Fuchsia* hybrids, Lady's Eardrops, have oval, mid-green leaves. Clusters of bell-shaped flowers, in combinations of white and red, purple and red, white and pink, appear from late spring to early autumn.

Softwood cuttings are taken in early spring or spring. Use 4in/10cm stem tip cuttings of overwintered plants. Remove the bottom pair of leaves and insert each cutting in a pot of half peat potting compost and half sand. Place in a plastic bag (supported by a wire hoop) and keep in the warmth, but out of direct sun. Cuttings should root in 4–6 weeks. When growing well, transplant to pots of peat potting compost.
Semi-ripe cuttings are taken in early autumn. Use 4in/10cm semi-ripe cuttings and treat them in the same way as softwood cuttings (above). When well established, remove from their plastic bags and overwinter at 50°F/10°C. Transplant them to pots of peat potting compost in early spring.
Seed is sown from winter to spring on the surface of peat seed compost in a pot or tray. Place in a propagator at 70–75°F/21–24°C. Germination takes anything from 3–13 weeks, so sow as early as possible for flowering in 5 to 6 months. What sort of flowers appear will be a gamble, but plants bearing the most desirable blooms can be overwintered and used to provide cuttings.

Hibiscus

SOFTWOOD CUTTINGS

Flowering shrub which grows to 6ft/1.8m. Temperatures required: summer, 60–65°F/16–18°C; winter, 55–60°F/13–16°C. *Hibiscus rosa-sinensis*, Rose of China, has oval, glossy, dark green leaves with serrated edges. Single or double funnel-shaped yellow, orange, red and pink flowers appear from early summer to early autumn.

Softwood cutting of *Hibiscus rosa-sinensis*

Softwood cuttings are taken in late spring or early summer. Use 4in/10cm stem cuttings with 3 or more pairs of leaves. Remove the bottom pair and plant in pots of half loam potting compost and half sand. Cover with a plastic bag, or put in a propagator, and keep at 65°F/18°C out of direct sun. When cuttings have rooted, new growth should appear. Take them out of the bag or propagator. When the pots are full of roots, pot-on into loam potting compost.

Hippeastrum

OFFSETS; SEED

Bulb producing leaves up to 2ft/60cm long and a flower stem of 20in/50cm. Temperatures required: summer, 55–60° F/13–16° C; winter, 50° F/10° C, with bulb stored dry. Once new growth begins, bring into temperature of 65–70° F/18–21° C. *Hippeastrum* hybrids have narrow, strap-shaped, bright green leaves usually preceded by a tall flower spike bearing trumpet-shaped white, pink, orange, red or bicoloured flowers.

Offsets are removed when bulbs are repotted at the first sign of new growth in winter or early spring. Remove bulb from the pot in which it has been stored dry after foliage has died down in autumn. Break compost away from the bulb to see if offsets are growing round the base. If so, they can be broken or cut away when 1in/2.5cm or more in diameter, with roots attached. Plant in a mixture of half loam potting compost and half sand. Pot-on 12 months later to individual pots using peat potting compost and treat as mature bulbs. It will be about 4 years before they reach flowering size.
Seed is sown in summer. Plant one seed to a 3in/7.5cm pot in a mixture of half loam potting compost and half sand. Cover seed to its own depth with the mixture and keep between 60–65° F/16–18° C. A propagator should not be necessary. Germination takes 10–14 days. Small bulbs will form from the seedlings and can be potted-on the following spring. Flowering size will be reached in 4 to 5 years.

Howea

SEED

Palm which grows to 7ft/2.1m. Temperatures required: summer, 55–65° F/13–18° C; winter, minimum 55–60° F/13–16° C. *Howea forsteriana*, Kentia Palm, has arching fronds of long, tapering, deeply-divided leaflets.

Seed is sown in spring. Soak for 24 hours in warm water or nick the seed coat with a sharp knife, taking care not to damage the inner tissue. Sow in a 3in/7.5cm pot of half loam potting compost and half sand. Place in a propagator at 80° F/27° C. Germination takes up to 6 months. When seedlings are about 3in/7.5cm tall, transplant to pots of loam potting compost. Return to the propagator, but gradually acclimatise to lower room temperatures over a period of 2 weeks.

Hoya

SOFTWOOD CUTTINGS

Flowering climber which grows to 6ft/1.8m and more, at the rate of about 18in/45cm each year. Temperatures required: summer, 60–65° F/16–18° C; winter, 50–55° F/10–13° C. *Hoya carnosa*, Wax Plant, has dark green, glossy leaves and white to pink flowers with a red star-shaped centre from early summer to summer. 'Variegata' has leaves edged white to pink.

Softwood cuttings are taken in late spring. Use stem tip cuttings about 4in/10cm long, severing them just below a pair of leaves. Remove the lowest pair and dip the stem end in hormone rooting powder. Insert several in a pot of half loam potting compost and half sand. Place in a propagator at 70° F/21° C. The cuttings should root in 8–10 weeks. Acclimatise cuttings to lower room temperatures over a period of 2 weeks before removing from the propagator. When roots have filled the pot, pot-on into a loam potting compost.

Hypoestes

SOFTWOOD CUTTINGS; SEED

Perennial foliage plant which grows to 12in/30cm. Temperatures required: summer, 65–70° F/18–21° C; winter, minimum 55–60° F/13–16° C. *Hypoestes sanguinolenta*, Pink Polka Dot Plant, has dark green, oval leaves, spotted and blotched with bright pink.

Softwood cuttings are taken from spring to early summer. Use stem cuttings, about 4in/10cm long, and remove the bottom pair of leaves. Insert several cuttings in a 3in/7.5cm pot of half peat potting compost and half sand. Place in a propagator at 70° F/21° C. Within 8 weeks, new growth should be seen, showing that cuttings have rooted. Remove from the propagator, and pot-on when roots have filled the pot, using peat potting compost. Pinch out growing tips regularly for bushy growth.
Seed is sown in early spring. Prepare a pot or tray with peat seed compost. Scatter seeds evenly over the surface of the compost and gently water them in. Do not cover with compost. Place in a propagator at 70°/21° C. When seedlings can be handled easily, thin out to 1in/2.5cm apart. Acclimatise them to room temperatures before removing from the propagator. When ready to transplant, grow several in a pot for the best effect.

Impatiens

SOFTWOOD CUTTINGS; SEED

Perennial flowering plant often treated as an annual and growing up to 15–24in/37.5–60cm. Temperatures required: summer, 60–65°F/16–18°C; winter, 60°F/16°C. *Impatiens wallerana*, Busy Lizzie, has elliptic, bright green, rubbery leaves. Red, orange, pink, white or bicoloured single or double flowers appear from late spring to early autumn. New Guinea hybrids grow to 2ft/60cm, having variegated leaves, creamy white to yellow, and larger flowers.

Softwood cuttings are taken in early summer, 4in/10cm long. Insert several in a pot of half peat potting compost and half sand. Cover with a plastic bag and keep in the warmth, but out of direct sun. A propagator is not necessary. Cuttings should root in 4–6 weeks. Pot-on, in peat potting compost, when roots have filled the pot.
Seed is sown in early spring. Prepare a pot or tray with peat seed compost. Scatter seed evenly over the surface and very lightly cover with compost. Place in a propagator at 70–75°F/21–24°C. Germination takes about 4 weeks. Remove from the propagator and thin seedlings to 1in/2.5cm. When 3in/7.5cm tall, transplant into pots of peat potting compost. Flowers after about 8 weeks.

Jacobina

SOFTWOOD CUTTINGS

Flowering shrub which grows 2–5ft/60–150cm tall. Temperatures required: summer, 60–65°F/16–18°C; winter, 55°F/13°C. *Jacobina carnea*, King's Crown, has oval, pointed, glossy dark green leaves with plumes of pink to red flowers in late summer.

Softwood cuttings are taken in spring. Use stem cuttings about 4in/10cm long, cutting just below a node. Remove the bottom pair of leaves, dip in hormone rooting powder and insert in a pot of half loam potting compost and half sand. Place in a propagator at 70°F/21°C. Rooting takes some 4 weeks. Remove from the

propagator and acclimatise to lower temperatures over a period of 2 weeks. When cuttings are well established, transplant into loam potting compost. For bushy growth, pinch out growing tips.

Lantana

SOFTWOOD CUTTINGS; SEED

Flowering climber which grows to 4ft/1.2m. Temperatures required: summer, 60–65°F/16–18°C; winter, 55–60°F/13–16°C, with the lower figure preferable. *Lantana camara*, Yellow Sage, has bright green, oval leaves and heads of tubular flowers changing from yellow through orange to red from late spring to autumn.

Softwood cuttings are taken in summer from non-flowering stems (this is essential). Cut just below a node and remove the bottom pair of leaves. Pot in half loam potting compost and half sand. Put in a propagator at 65°F/18°C and remove when cuttings have rooted, in about 4 weeks. When 4in/10cm high, pinch out growing tips for bushier growth. In spring the following year, pot-on in loam potting compost.
Seed is sown in early spring. Prepare a tray with loam seed compost. Soak seed for 24 hours in warm water, or scratch the surface with a sharp knife, taking care not to damage the inner tissue. Sow 1in/2.5cm apart and cover with compost to the depth of the seed. Place in a propagator at 70–75°F/21–24°C. Germination may take up to 3 months. When seedlings eventually reach about 3in/7.5cm, harden off over a period of 2 weeks. At 6in/15cm, pinch out growing tips for bushy growth.

Maranta

DIVISION; STEM CUTTINGS

Perennial foliage plant which grows to 12in/30cm, with a spread of 15in/37.5cm. Temperatures required: summer, 65–70°F/18–21°C; winter, 60–65°F/16–18°C. *Maranta leuconeura* 'Erythrophylla', Prayer Plant, has dark green oval leaves and crimson veins. 'Kerchoveana', Rabbit's Foot, has dark

brown blotches on the leaves, gradually turning dark green.

Division takes place in early spring. When potting-on, gently break away compost from the roots and divide the clump of leaves into 2 or 3 pieces. Plant each piece in a pot of peat potting compost. Division is the easiest propagation method for this plant.

Stem cutting of *Maranta leuconeura*

Stem cuttings are taken in early summer. Cut a piece of stem about 4in/10cm long, with 3 or more leaves. Remove from the base of the cutting any brown papery sheaths from which the leaves emerge. Plant in a pot of half peat and half sand. Cover with a plastic bag and keep warm, but not in direct sun. Rooting takes place in 6–8 weeks. When well established, transplant into peat potting compost.

Monstera

AIR LAYERING; SEED

Perennial climber which grows to 6ft/1.8m and more, with a similar spread.

Temperatures required: summer, 65–75°F/18–24°C; winter, similar, but tolerates temperatures down to 55°F/13°C. *Monstera deliciosa*, Swiss Cheese Plant, has large, dark green, heart-shaped leaves with many slashes and perforations. 'Variegata' has white to cream markings.

Air layering takes place in late spring and is suitable for a plant which has lost its lower leaves. Make an upwards slanting cut — not too deep — into the stem, 2in/5cm below the lowest surviving leaf. Dust the cut with hormone rooting powder and wedge it open with a matchstick. Cover the cut with moist sphagnum moss, shape it into a ball and secure with string. Cover the moss with clear plastic and seal it top and bottom with sticky tape. Water and feed the plant as usual. After about 10 weeks, the moss should be filled with roots. Remove the plastic and cut the stem just below the moss ball. Plant the rejuvenated monstera, along with the moss, in a pot of loam potting compost.

Seed is sown in spring. Prepare a pot or tray with loam seed compost. Sow evenly over the surface and cover with a very shallow layer of compost. Put in a propagator at 70–75°F/21–24°C. When young plants are 3in/7.5cm tall, transplant into loam potting compost. Slashes and perforations will not appear on the early leaves, but after 2 or 3 years these will be seen on the later, mature leaves.

Nertera

DIVISION; SEED

Perennial fruiting plant of creeping habit which forms a mound only 3in/7.5cm tall. Temperatures required: summer, 60°F/16°C, but will tolerate higher; winter, 50°F/10°C. *Nertera granadensis*, Bead Plant, has small, fleshy, oval leaves all but hidden by orange berries from summer through winter.

Division takes place when potting-on in spring. Remove plant from its pot and break off several small pieces with roots. Plant in a pot of two thirds peat potting compost and one third sand, arranging the pieces evenly so that they almost cover the surface of the compost. Because of the plant's creeping habit, any gaps will soon be hidden.

Seed is sown in early spring. Soak in warm water for 24 hours before sowing in a pot or tray of half peat seed compost and half sand. Scatter seeds evenly over the compost; do not cover, but gently water them into the compost. Place in a propagator at 65–70°F/18–21°C. Germination may take 6–12 weeks. When seedlings can be easily handled, transplant several into separate pots of two thirds peat potting compost and one third sand. Remove from the propagator and keep in a slightly shaded spot until they are growing well.

Peperomia

STEM CUTTINGS; LEAF CUTTINGS; DIVISION

Perennial foliage plant which grows 8–12in/20–30cm tall. Temperatures required: summer, 60–65°F/16–18°C; winter, as summer, but can tolerate down to 50°F/10°C. *Peperomia argyreia* (syn. *Peperomia sandersi*) Watermelon Plant, has fleshy, dark green, oval leaves, banded with silver, on stalks from compost level. *Peperomia caperata* 'Emerald Ripple' has heart-shaped leaves with purple and grey areas. *Peperomia magnoliifolia* 'Variegata' has mid-green leaves with yellow-green markings.

Stem cuttings of *Peperomia magnoliifolia* 'Variegata' are taken in spring, 4in/10cm long. Remove the bottom pair of leaves and insert several cuttings in a pot of half peat potting compost and half sand. Cover with a plastic bag and keep in the warmth in a slightly shaded spot. When cuttings have rooted, in about 6 weeks, take out of the bag. Pot-on using peat potting compost when the roots fill the pot.

Leaf cuttings of *Peperomia argyreia* and *Peperomia caperata* are taken in spring. Cut away several complete stalks with leaves. Trim the stalk to 1in/2.5cm and insert in a pot of half peat potting compost and half sand. The leaf must not touch the surface of the compost. Cover pot with a plastic bag and keep it warm. New growth should emerge from the base of the stem after 10 weeks or so. Remove the plastic bag and pot-on when roots have filled the pot.

Division takes place in spring when potting-on. Remove the plant from the pot and gently break away compost from the roots. Pull these apart, dividing into 2–3 pieces. Plant each piece in a pot of peat potting compost and keep out of direct sun until the plant has settled down.

PINCHING OUT

Gardeners are frequently exhorted to pinch out the growing tip of a young plant's stem, to encourage bushiness. This operation (also known as stopping) is simple. Not all plants have growing tips, but there is no mistaking it in those that do, and no difficulty in removing it between thumb nail and finger. It is also effective. But why?

The active bud on the end of a stem is the apical (or terminal) bud. In some mysterious way it inhibits development of lateral buds lower down the stem, while the plant goes on growing to its natural height. Once the apical bud is removed, the lower dormant buds can start into growth

and produce lateral stems. The growing tips of these lateral stems can in turn be pinched out later.

Philodendron

SOFTWOOD CUTTINGS;
LEAF BUD CUTTINGS; SEED

Shrubs of climbing and erect types. Climbers grow to 8ft/2.4m and more; erect types to 1–5ft/30cm–1.5m, with a spread of 3ft/90cm. Temperatures required: summer, 65–70°F/18–21°C; winter, 60–65°F/16–18°C, but will stand temperatures down to 55°F/13°C.

Climbing species include *Philodendron scandens*, Heartleaf Philodendron, with dark green, glossy, heart-shaped leaves. *Philodendron andreanum*, Velour Philodendron, has elongated heart-shaped, velvety leaves with purple undersides; *Philodendron erubescens*, Blushing Philodendron, has arrow-shaped leaves, pink when young, ageing to dark green. *Philodendron panduraeforme*, Fiddle Leaf, has dark green, fiddle-shaped leaves.

Softwood cuttings of climbers are taken in late spring or early summer. For plants with manageable-sized young leaves (*P. scandens* and *andreanum*), take 4in/10cm stem tip cuttings with at least 3 leaves. Make the cut just below a node and remove the lowest leaf. Insert in a pot of half peat potting compost and half sand and put in a propagator at 70–75°F/21–24°C. Rooting takes place within 4–6 weeks. Harden off over a period of 2 weeks before removing from the propagator. When the pot is full of roots, pot-on into peat potting compost.

Leaf bud cuttings are taken in late spring or early summer from species with larger leaves. Cut off the top of a stem with at least 2 or 3 leaves. Then divide it

so that each piece has one leaf and a length of stem 1in/2.5cm above the leaf stalk and 2in/5cm below it. Fill a pot with half peat potting compost and half sand. Insert the longer end of the stem into the compost so that the point where the leaf stalk joins the stem is at compost level. If the cutting is not too big, place in a propagator at 70–75°F/21–24°C. Support the leaf with a cane if necessary. Alternatively, enclose in a plastic bag and keep the temperature as near to 70–75°F/21–24°C as possible. Rooting will take several weeks, after which new growth should appear from the leaf axil — the point where the leaf stalk joins the stem. When growing well, remove from the propagator or bag.

Erect growing species include *Philodendron bipinnatifidum*, with lobed leaves. *Philodendron selloum* has similar, but much larger, leaves and deeper dissections. *Philodendron wendlandii* is more compact, with lance-shaped leaves in rosette formation.

Seed is sown in spring in peat seed compost. Do not cover the seed of *Philodendron bipinnatifidum*, but cover the others with compost to the depth of the seed. Place in a propagator and keep at 80–85°F/27–30°C. When seedlings are about 3in/7.5cm high, and growing well, transplant into pots of peat potting compost. Gradually harden off.

Phoenix

SEED; SUCKERS

Palm which grows to 6ft/1.8m, with a spread of 3ft/90cm. Temperatures required: summer, 65–70°F/18–21°C; winter, 55–60°F/13–16°C. *Phoenix canariensis*, Canary Date Palm, has erect fronds with opposite pairs of tapering pinnae. The fronds of *Phoenix roebelenii* are more arching.

Seed is sown in spring. Fill a 3in/7.5cm pot with loam seed compost. Soak the seed in warm water for 24 hours, or cut or scrape the hard coat with a sharp knife, taking care not to damage the inner tissue. Plant seeds to their own depth in the compost and place in a propagator at 75–80°F/24–27°C. Germination is erratic and can take 2–6 months. When a seedling

is about 3in/7.5cm high, transplant into a mixture of two thirds loam potting compost and one third sand. Gradually harden off. Growth will be very slow.

Suckers of *Phoenix roebelenii*, sometimes thrown up from the base of the plant, are taken in spring. Remove the plant from its pot and gently dislodge compost from the roots. Break away or cut the sucker from the parent plant, taking enough roots to support it. Plant in two thirds loam potting compost and one third sand. Keep warm in a slightly shaded spot until the sucker is growing well.

Pilea

SOFTWOOD CUTTINGS; SEED

Perennial foliage plant, reaching 6–12in/15–30cm. Temperatures required: summer, 55–60°F/13–16°C; winter, as summer, but 50°F/10°C is tolerated. *Pilea cadieri*, Aluminium Plant, has oval, dark green leaves with a silvery quilted effect. Other species can be propagated in the same ways.

Softwood cuttings are taken in late

KEEP IT CLEAN

Dirty leaves deprive a plant of sunlight and the air it breathes, both of which it needs to manufacture food. Wipe dust from large glossy leaves and then sponge them, both top and underside, with a cloth dampened in tepid water. Use nothing else but water. Mist spray small glossy leaves.

spring. It is best to take cuttings annually since pileas often lose their lower leaves after one year. Use 3in/7.5cm stem cuttings. Remove the bottom pair of leaves, dip the stem ends in hormone rooting powder and insert several in a pot of two thirds peat potting compost and one third sand. Cover with a plastic bag and keep around 65°F/18°C in a slightly shaded spot. A heated propagator is not necessary. Rooting takes 4–6 weeks. When cuttings are growing well, remove from the plastic bag and later pot-on in peat potting compost.

Seed is sown in spring. Fill a pot or tray with peat seed compost. Scatter seeds evenly over the surface and gently water them into the compost. Place in a propagator at 70°F/21°C. Thin seedlings to 1in/2.5cm and when 3in/7.5cm high, transplant to individual pots in peat potting compost. Remove from the propagator.

Rhoeo

OFFSETS; SEED

Perennial foliage plant with leaves up to 12in/30cm long. Temperatures required: summer, near to 60°F/16°C, but higher will be tolerated; winter, 50°F/10°C. *Rhoeo spathacea*, Boat Lily, has a rosette of lance-shaped, olive green leaves with purple undersides. White flowers emerge from purple boat-shaped bracts in late spring to summer.

Offsets are removed when repotting in spring. Small replicas of the parent appear at the base of the plant. Those which are at least 4in/10cm tall, and have several leaves, can be detached. Remove the plant from its pot and ease compost away from the roots. Break away the offset, but only if there are enough roots to support it as a separate plant. If large enough to detach, plant the offset in peat potting compost and keep warm in a shaded spot until growing well.

Seed is sown in spring. Fill a pot or tray with peat seed compost. Soak seeds in warm water for 2 hours and then sow evenly, covering to their own depth. Place in a propagator at 70°F/21°C. Germination takes 8 weeks. Remove seedlings from the propagator when they

are 3in/7.5cm tall, having gradually acclimatised them to a somewhat lower temperature. At 4–5/10–12.5cm, move into pots of peat potting compost.

Rhoicissus

SOFTWOOD CUTTINGS; SEED

Climbing plant which grows to 6ft/1.8m. Temperatures required: summer, 55–65°F/13–18°C; winter, 50°F/10°C. *Rhoicissus capensis*, Cape Grape, has rounded to heart-shaped bright green, glossy leaves with toothed edges.

Softwood cuttings are taken in spring. From stem tip cuttings about 4in/10cm long, remove the bottom leaves and insert in a mixture of half loam potting compost and half sand. Cover with a plastic bag and keep in the warmth in a slightly shaded spot. Cuttings should root in about 6 weeks. When they are growing well, remove the plastic bag.

Seed is sown in spring. Fill a pot or tray with loam seed compost. Sow evenly over the surface and cover seeds with compost to their own depth. Place in a propagator at 70°F/21°C. When seedlings have reached 3in/7.5cm, remove from the propagator and transplant to individual pots of loam potting compost.

Saintpaulia

DIVISION; LEAF CUTTINGS; SEED

Flowering perennial which grows to 6in/15cm, with a spread to 9in/22.5cm. Temperatures required: summer, 60–70°F/16–21°C; winter, about the same. *Saintpaulia ionantha*, African Violet, has heart-shaped, velvety leaves carried on stalks from compost level. Single and double white, pink, red, purple or bicoloured flowers appear throughout the year.

Division takes place in spring. A pot may hold 2 or more rosettes of leaves and when these become too big for the container, each can be potted up singly. Remove the plant from its pot gently and ease compost away from the roots. While doing this, the individual rosettes come apart easily. Pot each one in peat potting compost.

Leaf cuttings of *Saintpaulia ionantha*

Leaf cuttings are planted in spring. Break off a few leaves with their stems. Cut the stems down to 1in/2.5cm long and dip the base in hormone rooting powder. Prepare a pot of peat potting compost, topped with a layer of sharp sand. Insert cuttings at an angle of 45° and to a depth of 0.5in/1.25cm. Leaves should not touch each other or rest on the sand. Place in a propagator at 70–75°F/21–24°C. Plantlets appear from the base of the stalks, but it may take 10 weeks or more. When growing well, acclimatise plantlets gradually to lower temperatures outside the propagator. When about 1in/2.5cm tall, detach them from the leaf and transplant to individual pots of peat potting compost.

Seed is sown in spring. Fill a pot or tray with peat seed compost. Sow evenly, but do not cover seeds with compost; gently water them in. Place in a propagator at 70°F/21°C. Germination takes 8 weeks. If necessary, thin seedlings to 1in/2.5cm. When about 1in/2.5cm tall, accustom them over 2 weeks to lower room temperatures and transplant to individual pots of peat potting compost.

Sansevieria

DIVISION (ROOTED SUCKERS);
LEAF CUTTINGS (SECTIONS)

Perennial foliage plant with leaves
growing to 18in/45cm plus. Temperatures
required: summer, 60–70°F/16–21°C;
winter, down to 50°F/10°C is tolerated.
Sansevieria trifasciata 'Laurentii' has
narrow, sword-shaped, mid-green leaves
with darker horizontal bands and yellow
edges. Other sansevierias form low-
growing rosettes.

Division (rooted suckers) takes place in
spring when repotting crowded plants.
Remove from the pot and break away
compost from the roots to reveal the
rhizome from which the leaves grow. Cut
away rosettes of leaves and plant in pots of
loam potting compost.

Leaf cuttings of *Sansevieria trifasciata*
'Laurentii'

Leaf cuttings (sections) are taken in
early summer. This method produces
plain green plants, without any
variegations, so may not be thought worth

the effort. Cut away a complete leaf and
divide it into 2in/5cm sections. As you cut
each one, nick the base end of the section
as a reminder of which way up to plant it.
Insert the bottom, nicked, edge into the
potting compost mixture of half loam and
half sand. Cover with a plastic bag and
keep in the warmth. New growth appears
from the base of the cuttings, but may
take months. When growing well, remove
the plastic bag.

Saxifraga

LAYERING PLANTLETS

Perennial foliage plant which grows to
8in/20cm, with runners up to 24in/60cm.
Temperatures required: summer,
50–60°F/10–16°C; winter, 45°F/7°C.
Saxifraga stolonifera, Mother of
Thousands, has rounded, slightly hairy,
dark green leaves with silver veins and red
undersides. Runners bear small replicas of
the parent plant.

Layering plantlets takes place in spring.
Pin down the plantlets, still attached to
the runners, in pots of half loam potting
compost and half sand. Rooting takes
about 6–8 weeks. When plantlets are well
established, sever from the runners.
About 6 weeks later repot into loam
potting compost.

Scindapsus

SOFTWOOD CUTTINGS

Evergreen climber growing to 6ft/1.8m or
more. Temperatures required: summer,
65–70°F/18–21°C; winter, 60°F/16°C is
best, with minimum 55°F/13°C.
Scindapsus aureus, Devil's Ivy, has green
leaves splashed with yellow; 'Marble
Queen' has creamy white markings and
'Tricolor' white and yellow.

Softwood cuttings are taken in late
spring. Make a cut below a node on
4in/10cm stem cuttings. Remove the
bottom leaf and dip the stem end in
hormone rooting powder. Insert several
cuttings in a pot of half peat potting
compost and half sand. Place in a
propagator at 65–70°F/18–21°C. New
growth appears in 6–8 weeks, indicating

that cuttings have rooted. Gradually
harden off to lower room temperatures
and pot-on later using peat potting
compost.

Setcreasea

DIVISION; SOFTWOOD CUTTINGS

Perennial foliage plant with trailing stems
reaching 12–18in/30–45cm.
Temperatures required: summer,
65–70°F/18–21°C; winter,
60–65°F/16–18°C. *Setcreasea purpurea*,
Purple Heart, has lance-shaped purple
leaves on purple stems. Small, deep pink
flowers appear during summer months.

Division takes place in spring when
repotting large plants. Remove plant from
its pot and gently break away compost
from the roots. Pull the root ball apart
into 2 pieces and plant them in peat
potting compost.
Softwood cuttings are taken in late
spring. Remove the lowest pair of leaves
from stem tip cuttings about 4in/10cm
long. Insert cuttings in peat potting
compost and keep warm in a slightly
shaded spot. They need neither a
propagator nor plastic bag. Cuttings
should root easily within 4 weeks.

Sinningia

LEAF CUTTINGS; SEED

Perennial flowering plant which grows
3–12in/7.5cm–30cm high. Temperatures
required: summer, 60–70°F/16–21°C;
winter (when leaves die down and tubers
are stored), 55°F/13°C. *Sinningia speciosa*
hybrids, Gloxinias, have rosettes of oval,
velvety, mid-green leaves. Violet, dark
red, pink, white and bicoloured trumpet-
shaped flowers bloom from late spring to
early autumn. *Sinningia pusilla* hybrids
are miniature plants, no more than
3in/7.5cm high, with narrow, trumpet-
shaped flowers.

Leaf cuttings of *Sinningia speciosa*
hybrids are taken in spring. Fill a tray or
pot with moist peat potting compost, with
a thin layer of moist sharp sand on top.
Remove a complete leaf and stalk. Trim
off the stalk and turn the leaf upside

down. With a sharp knife, make cuts into the main central vein and the veins which join it, but take care not to cut right through. Place the leaf on the sand, right side up, and pin down with loops of wire so that the cut veins are in close contact with the surface. Cover with a plastic bag and keep warm. After about 10 weeks, plantlets should appear from the cut veins. When they have several leaves and can easily be handled, gently pull away from the leaf and plant in peat potting compost.

Seed of both types of sinningia hybrids is sown in spring. Fill a tray or pot with peat seed compost. Sow evenly over the surface and gently water seeds in. Do not cover with compost as they need light to germinate. Put in a propagator at 65–70°F/18–21°C. Germination takes about 4 weeks. When seedlings have several leaves and can be easily handled, transplant to individual pots of peat potting compost. Four weeks later begin to acclimatise to room temperatures, before removing from the propagator altogether after 2 weeks.

Sonerila

BASAL CUTTINGS

Perennial foliage and flowering plant which grows to 8in/20cm. Temperatures required: summer, 65–75°F/18–24°C; winter, 65–70°F/18–21°C. *Sonerila margaritacea*, Frosted Sonerila, has dark green, oval, pointed leaves with silvery white spots. Rosy pink flowers appear from late spring to late summer.

Basal cuttings are taken in spring. Use cuttings some 3in/7.5cm long, thrown up about the base of the plant. Remove the bottom pair of leaves, dip the base of the stem in hormone rooting powder and insert in a mixture of half peat potting compost and half sand. Place in a propagator at 70–75°F/21–24°C. Rooting takes about 6 weeks. When cuttings are 2in/5cm high, acclimatise gradually to lower room temperatures over a period of 2 weeks and then remove from the propagator altogether. After a further 2–3 months, pot-on, using a mixture of peat potting compost and leaf mould.

Spathiphyllum

DIVISION

Perennial foliage and flowering plant which grows 12–24in/30–60cm. Temperatures required: summer, 65–70°F/18–21°C; winter, 60–65°F/16–18°C. *Spathiphyllum wallisii*, Peace Lily, has glossy, lance-shaped leaves on long stalks. In late spring, flowers are followed by white spathes surrounding yellow spadices.

Division takes place in spring when potting-on. Remove plant from its pot and gently ease away compost from the roots to reveal the rhizome. Pull the rhizome apart into 2 or 3 pieces, each with plenty of roots to support the leaves. Plant each piece in peat potting compost and keep warm in a slightly shaded spot.

Stephanotis

SEMI-RIPE CUTTINGS; SEED

Climber which grows to 10ft/3m. Temperatures required: summer, 65–70°F/18–21°C; winter, 55°F/13°C. *Stephanotis floribunda*, Madagascar Jasmine, has leathery, shiny, dark-green leaves, with clusters of white waxy flowers in late spring and early summer.

Semi-ripe cuttings are taken in late spring. Choose 4in/10cm cuttings from lateral non-flowering stems of the

Semi-ripe cuttings of *Stephanotis floribunda*

previous year's growth. Remove the bottom pair of leaves and dip the stem end in hormone rooting powder. Plant several in a pot of half peat potting compost and half sand. Place in a propagator at 70°F/21°C. Rooting takes about 10–12 weeks.

Seed is sown in early spring. Fill a pot with peat seed compost. Space the seeds evenly and cover to their own depth with compost. Place in a propagator at 75–80°F/24–27°C. Germination takes up to 12 weeks. When seedlings are growing well, remove from propagator. Three weeks later repot in peat potting compost.

Strelitzia

DIVISION; SEED

Perennial flowering plant which grows to 4ft/1.2m. Temperatures required: summer, 60–70°F/16–21°C; winter, 55–60°F/13–16°C. *Strelitzia regina*, Bird of Paradise, has spear-shaped leaves on long stalks. From early to late spring, large blue and orange petalled flowers appear from a red-edged green bract on a 3ft/90cm stem.

Division takes place in spring when repotting a plant with several clumps of leaves. Remove plant from its pot and gently break away compost from the roots. Split the clumps of leaves, making certain that each clump has plenty of roots to sustain it. Plant separately in a pot of peat potting compost. Keep warm, but out of direct sun, until plants recover.

Seed is sown in early spring, but the seeds take months to germinate and it will be 7 or more years before the plant flowers. Soak seeds in warm water for 24 hours before sowing, or chip a little of the hard surface with a sharp knife, taking great care not to damage the inner tissue. Prepare a pot of half peat potting compost and half sand. Cover seeds with compost to their own depth and place in a propagator at 70–75°F/21–24°C. Germination can take up to 6 months. When seedlings have reached about 3in/7.5cm, acclimatise them gradually to lower room temperatures before removing from the propagator altogether, around 2 weeks later. After 2 months, transplant to pots of peat potting compost.

Indoor Plants

Streptocarpus

LEAF CUTTINGS; SEED

Perennial flowering plant which grows to 12in/30cm. Temperatures required: summer, 65–70°F/18–21°C; winter, 55–60°F/13–16°C. *Streptocarpus hybrids*, Cape Primrose, have rosettes of strap-shaped, wrinkled, bright green leaves. Funnel-shaped flowers in many colours appear from late spring to early autumn.

Leaf cuttings are taken in spring. Remove a complete leaf and turn it upside down. Make a cut along the length of the main vein, taking care not to cut right through. Fill a pot with peat potting compost, with a layer of moist sharp sand on top. Place the leaf, top side uppermost, on the surface of the sand so that the cut vein is in close contact with it. Pin down with wire loops. Cover the pot with a plastic bag and keep warm. Expect plantlets to grow in about 6 weeks. When about 3in/7.5cm tall, detach from the leaf, along with the roots, and plant in individual pots of peat compost.

Seed is sown in spring. Fill a pot or tray with peat seed compost. Scatter seeds evenly over the surface and gently water into the compost. Do not cover, since they need light to germinate. Cover the pot or tray with clear plastic and keep in a warm place; a heated propagator is not necessary. Germination takes 4–6 weeks. Thin seedlings to 1in/2.5cm, and when 3in/7.5cm tall, transplant to pots of peat potting compost.

Tolmiea

LEAF PLANTLETS

Perennial foliage plant which grows to 12in/30cm. Temperatures required: summer, 55°F/13°C preferable, but high temperatures tolerated; winter, 50°F/10°C. *Tolmiea menziesii*, Piggyback Plant, has hairy, heart-shaped, bright green leaves, carrying plantlets where the leaf stalk joins the leaf.

Leaf plantlets are removed in late spring or early summer. Take leaves with well-developed plantlets, complete with their stem. Trim stems to 2in/5cm and insert in a pot of half peat potting compost and half sand. They should be deep enough for the point where the plantlet joins the stem to rest on the surface of the compost. Rooting takes about 4 weeks. When growing well, repot in peat potting compost, along with the leaf stalk, which will eventually die.

Tradescantia

SOFTWOOD CUTTINGS

Perennial foliage plant with trailing stems growing to 2ft/60cm. Temperatures required: summer, 65°F/18°C; winter, minimum 50°F/10°C. *Tradescantia fluminensis*, Wandering Jew, has oval, bright green leaves with pale purple undersides. Other species and varieties have different colour combinations.

Softwood cuttings are taken from spring to summer. Remove the lower leaves of 4in/10cm stem tip cuttings and insert several in a mixture of half peat potting compost and half sand. Keep in the warmth in a slightly shaded spot; they need no covering. Rooting takes 2–3 weeks. When growing well, repot in peat potting compost.

Yucca

OFFSETS; SEED; STEM SECTIONS

Perennial foliage plant which grows to 6ft/1.8m. Temperatures required: summer, 55–65°F/13–18°C; winter, 45–50°F/7–10°C. *Yucca aloifolia*, Spanish Bayonet, has a tall spiky trunk, topped with rosettes of sword-shaped, dark green leaves, edged with fine teeth. *Yucca elephantipes* is similar, but the edges are toothless.

Offsets are removed in spring when potting-on. These appear at the base of the stem and can be detached when they have reached 12in/30cm — not before. Take the plant from its pot, remove some compost from the roots and cut away the offset, along with plenty of roots. Plant in a pot of loam potting compost. Keep out of direct sun until it is growing well.
Seed is sown in spring. Fill a pot or tray with seed compost. Sow evenly over the surface and cover seeds to their depth with compost. Place in a propagator at 70–75°F/21–24°C. Germination is erratic, taking anything from 2 to 12 months. When seedlings are about 3in/7.5cm tall, accustom them to lower room temperatures before removing from the propagator. Three months later, transplant into pots of loam potting compost.
Stem sections of *Yucca elephantipes* are taken in spring. The method can be used to produce yucca cane plants from an ageing yucca which has lost its lower leaves, but it will be a long time before the new plant is well developed. Cut the stem into pieces, the length depending on how tall the new plant is to be, for the stem itself will never grow any taller. (And a tall stem will need a tall propagator.) Dip the base of each stem in hormone rooting powder and insert, bottom end down, in potting compost of half peat and half sand. Place in a propagator at 75°F/24°C. Growth should eventually appear from the dormant buds on the stem, showing that the stem has rooted. When rosettes of leaves are about 3in/7.5cm tall, take the plant from the propagator. Three months later pot-on, using loam potting compost.

Zebrina

SOFTWOOD CUTTINGS

Perennial foliage plant with trailing stems growing to 15in/37.5cm. Temperatures required: summer, 65°F/18°C; winter, minimum 55°F/13°C. *Zebrina pendula* 'Quadricolor', Wandering Jew, has oval green leaves striped cream, pink and silver, with purple undersides.

Softwood cuttings are taken from spring to summer. Remove the lower leaves from stem tip cuttings about 4in/10cm long. Insert several in a pot of half peat potting compost and half sand. Keep in the warmth, but out of direct sun. Rooting takes place in about 4 weeks. When the pot is filled with roots, pot-on into peat potting compost to which a little sand has been added.

Bromeliads

Bromeliads are curious tropical or subtropical plants which grow as rosettes. Some are epiphytic. Their minimal roots are used mainly for clinging to the trunks of trees, while food and water are taken in through the leaves. Other bromeliads live on the ground. Almost all those grown as houseplants are epiphytes, but there is no difficulty in growing them in a pot. The one widely grown terrestrial bromeliad is the pineapple.

Bromeliads are mostly grown for their leaves and brilliantly coloured bracts, not for their flowers. The leaves have other functions besides absorbing food and water. Their colour attracts pollinating insects and many of them grow to form a cup. This collects not only potential plant food floating in the air, but also water, so acting as a reservoir for the plant to use in times of drought. In the wild, the plant cup is replenished by rain and dew, but indoors this is a routine task for the cultivator. Houseplant bromeliads can be supplied with food in the form of fertilizers.

Another curious characteristic of many bromeliads is that after the rosette has flowered once only, it dies, leaving behind offsets to carry on the life of the plant. It may take from two to twenty years before the offspring flower and then in turn die, but most of the popular bromeliads grown as houseplants will flower in two to three years. Propagation is almost entirely achieved by removing offsets, although three of the bromeliads in the following plant entries — aechmea, billbergia and neoregelia — offer the alternative of growing plants from seed.

Offsets should be allowed to reach a good size — between 3–6in/7.5cm–15cm — before they are cut away, taking plenty of roots. Details of the appropriate composts and temperatures needed for rooting are given in the following entries.

Bromeliads such as *Aechmea* produce offsets at the base before the old plant dies away.

Bromeliads

Aechmea

OFFSETS; SEED

Perennial foliage and flowering plant with leaves up to 2ft/60cm long and an overall spread of 3ft/90cm. Temperatures required: summer, 70°/21°C; winter, 55–60°F/13–16°C. *Aechmea fasciata* (syn. *Aechmea* or *Billbergia rhodocyanea*), Urn Plant, has a rosette of strap-shaped, grey-green leaves banded with silvery markings. In summer months, blue flowers surrounded by pink bracts emerge on a flower spike from the rosette.

Removing an offset from *Aechmea fasciata*

Offsets are removed in late spring or early summer. When the plant has finished flowering, it throws out offsets from the base and then dies. Wait until the offsets are at least 6in/15cm tall, with plenty of leaves, before detaching them. Remove the plant from its pot and gently break away compost around the offset, taking care not to damage the delicate roots. Cut it from the parent, taking plenty of roots, and plant in a mixture of loam, peat and leaf mould. Keep in a

warm place out of direct sun until well established. Offsets may not flower for 2 or 3 years.

Seed is sown in spring. Fill a pot or tray with peat seed compost and sow evenly over the surface. Gently water seeds into the compost, but do not cover as they need light to germinate, which takes up to 3 months. Place in a propagator at 70°F/21°C. When about 3in/7.5cm tall, transplant seedlings to individual pots of loam, peat and leaf mould. Gradually acclimatise to lower room temperatures over a period of 2 weeks before removing from the propagator. It will be 4–5 years before plants flower.

Ananas

OFFSETS

Perennial foliage, flowering and fruiting plant with leaves up to 3ft/90cm long and a spread to 4ft/1.2m. Temperatures required: summer, 65–75°F/18–24°C; winter, minimum 60°F/16°C and higher if possible. *Ananas comosus variegatus*, Pineapple, has a rosette of sword-shaped, grey-green leaves with white to yellow margins and sharply serrated edges. A spike of purple flowers emerges from it in spring, followed by the pineapple fruit.

Offsets are removed in late spring. The parent plant produces offsets at its base before it dies down after flowering and fruiting. Allow the offset to grow to 6in/15cm before detaching it. Remove plant from its pot and gently ease compost away from the offset. Cut it away with any roots that have developed and plant in a mixture of half peat potting compost and half sand. Place in a propagator at 75°F/24°C. Leave for 8–10 weeks until well established. After a further 3 months, pot-on using a lime-free mixture of loam, peat and leaf mould.

Billbergia

DIVISION (OFFSETS); SEED

Perennial foliage and flowering plant with leaves up to 20in/50cm long and a spread to 3ft/90cm. Temperatures required: summer, 60–65°F/16–18°C; winter, similar, but temperatures down to

50°F/10°C are tolerated. *Billbergia nutans*, Queen's Tears, has rosettes of dark green, arching leaves. Green and blue drooping flowers with pink bracts may appear at any time of the year.

Division (offsets) takes place in spring when repotting overcrowded clumps of leaf rosettes. Offsets should be at least 6in/15cm tall before they are detached. Remove plant from its pot and break away compost from the roots. Either cut away 1 or 2 offsets or split the whole plant in 2 pieces. Plant in peat potting compost and keep warm, out of direct sun, until growing well. Individual offsets may need some support.

Seed is sown in spring. Fill a pot or tray with peat seed compost. Space out seed evenly over the surface; this saves later thinning out. Gently water the seed into the compost but do not cover it. Place in a propagator at 70°F/21°C. Germination takes 8–12 weeks. When seedlings are about 3in/7.5cm tall, transplant to individual pots of peat potting compost. Acclimatise plants to lower room temperatures over a period of 2 weeks before removing completely from the propagator.

Cryptanthus

OFFSETS

Perennial foliage plant with leaves 4–12in/10–30cm long. Temperatures required: summer, 65–75°F/18–24°C; winter, minimum of 60°F/16°C. *Cryptanthus bivittatus*, Earth Star, has olive green, undulate leaves striped with darker green and tinged red to pink; *Cryptanthus acaulis* has small rosettes of mid-green leaves with spiny edges; *Cryptanthus bromeleoides* 'Tricolor', Rainbow Star, has bronze-green and ivory striped leaves edged with red.

Offsets are removed when repotting in spring. Offsets from the base of the plant can be detached if they have reached at least 2in/5cm across, with several leaves. Remove plant from its pot and gently ease compost from the selected offset. Break it away, or cut with a sharp knife if necessary. Plant in a mixture of peat potting compost and sand and place in a propagator at 70°F/21°C. After about 12 weeks, the offset should have enough roots for healthy growth. Remove from the propagator, after it has been acclimatised to lower room temperatures over 2 weeks. Two months later pot-on, using peat potting compost.

Guzmania

OFFSETS

Perennial foliage and flowering plant with leaves 12–18in/30–45cm long. Temperatures required: summer, 65–70°F/18–21°C; winter, 60–65°F/16–18°C. *Guzmania lingulata*, Scarlet Star, has a rosette of sword-shaped, narrow green leaves often tinged red. Bright red bracts surrounding yellow flowers grow from the middle of the rosette of leaves and remain from autumn through to winter. *Guzmania lingulata* 'Minor' has smaller leaves but is similar in appearance. *Guzmania zahnii* has translucent leaves striped red on upper and lower surfaces.

Offsets are removed in spring. When plants produce bracts and flowers, offsets will also grow. Wait until offsets are at least 3in/7.5cm tall, with several leaves, before detaching them. Remove plant from its pot and break away compost from the offsets to see where they join the parent. Cut them away with the roots and plant in either peat potting compost or a mixture of loam, peat and leaf mould. Keep warm, but out of direct sun, until growing well.

Neoregelia

OFFSETS; SEED

Perennial foliage and flowering plant with leaves up to 12in/30cm and a spread to 18in/45cm. Temperatures required: summer, 65–70°/18–21°C; winter, 55–60°F/13–16°C. *Neoregelia carolinae* 'Tricolor', Blushing Bromeliad, has sword-shaped, shiny, bright green leaves striped yellow. Before the plant is about to flower, the middle of the rosette turns bright red and the leaves are tinged pink. Small white flowers grow from the middle of the rosette.

Offsets are removed in spring. They will be produced around the base of the plant and should not be detached until they grow about 6in/15cm tall, with plenty of leaves. Gently break compost from the offset and cut it away with enough roots for its support. Plant in a pot of half peat potting compost and half sand and cover with a plastic bag. Keep warm, but out of direct sun. After about 10 weeks, when the offset is well established, remove the bag and repot in peat potting compost or a mixture of peat, loam and leaf mould.
Seed is sown in spring. Fill a pot or tray with peat seed compost and sow evenly over the surface. Gently water seeds into the compost but do not cover. Place in a propagator at 70°F/21°C. Germination can take 8–12 weeks. When seedlings are 3in/7.5cm tall, transplant to pots of peat potting compost or a mixture of loam, peat and leaf mould. Accustom seedlings to lower room temperatures over 2 weeks before removing from the propagator.

Tillandsia

OFFSETS; STEM PIECES

Tillandsia lindeniana

Perennial foliage and flowering plant with leaves growing to 15in/37.5cm. Temperatures required: summer, 60–65°F/16–18°C; winter, minimum 55°F/13°C. *Tillandsia lindeniana*, Blue-flowered Torch, has a rosette of grey-green leaves tinged purple on the undersides. Blue flowers with white throats emerge in summer from pink overlapping bracts in fan-shaped formation. *Tillandsia cyanea* is similar, but with reddish brown stripes on leaf undersides. *Tillandsia usneoides*, Spanish Moss, is a mass of wiry grey-green stems.

Offsets are removed in spring. They grow from the base of the parent plant and should not be detached until they are about 6in/15cm, with plenty of leaves. Remove plant from its pot and gently break away compost from the offset to discover where it is attached to the main plant. Cut it away with as many roots as possible and plant in a pot of half peat potting compost and half sand. Cover with a plastic bag and keep in the warmth, out of direct sun. After about 3 months, when the plant is growing well, remove the plastic bag and repot in peat potting compost or a mixture of peat, loam and leaf mould.
Stem pieces of *Tillandsia usneoides* are removed in late spring. Take several stems and wire them with plastic-coated wire to a small piece of cork or bark. This plant is not grown in compost.

Vriesea

OFFSETS

Perennial foliage and flowering plant with leaves up to 12in/30cm long and a spread to 20in/50cm. Temperatures required: summer, 65–75°F/18–24°C; winter, minimum 65°F/18°C. *Vriesea splendens*, Flaming Sword, has a rosette of strap-shaped, dark green leaves banded with brownish red. The spike of yellow flowers emerging from overlapping bright red bracts appears in summer.

Offsets are removed in spring when repotting. Not until these are at least 6in/15cm tall, with several leaves, should they be detached. Remove the plant from its pot and gently ease away the compost from the offsets. With a sharp knife, cut away the offset, with roots attached, from the parent. Plant in a mixture of sphagnum moss, sand and peat. Enclose the pot in a plastic bag and keep in the warmth, but not in direct sun. After about 8 weeks, when the plant is growing well, remove the plastic bag. Some offsets may grow at leaf axils, the point where the leaf joins the main stem. These should not be removed but left to take over from the main plant, which will eventually die

Cacti and Succulents

Succulent plants — and that includes cacti — form a weird and wonderful class of their own. Most plants need leaves to live; the sun that falls on them has a vital part in manufacturing their food. But plants also lose water through the stomata (pores) of the leaves, and at a great rate when exposed to hot sun. Unless the soil holds a reserve of water from which the roots can replenish these losses, the plant will die.

Cacti have overcome this problem by doing without leaves. Instead they have stems which do the work of leaves and which can also store water. Cacti from desert areas of the world use up their water reserves in times of drought and the stems shrivel. When the rains come, the stems swell again as the reserves are replaced.

Epiphytic jungle cacti have also developed swollen stems, not because they suffer from drought but because their few roots grow in the crevices of trees. They cannot draw up water from the soil, but instead use the store of water in their stems. When they can, they build up new reserves from rain or dew.

Cacti are stem succulents; other succulents are more likely to store water in their fleshy leaves.

RAISING FROM SEED

Cacti and succulents are propagated by cuttings and offsets, as well as from seed. Raising from seed is comparatively simple, but as with most seeds other than annuals, patience is needed for the seedling to become a mature flowering plant — the fastest cacti take at least two years. All desert and jungle cacti, and succulents, need much the same treatment when

Cacti and succulents grow in a varied range of shapes and sizes. Many can be propagated from portions of stem or from offsets.

Cacti and Succulents

Sowing seed

Cover a seed tray base with gravel or perlite. Top with compost and a thin layer of sand.

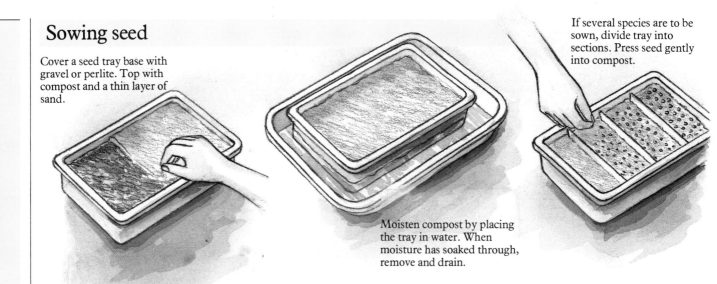

Moisten compost by placing the tray in water. When moisture has soaked through, remove and drain.

If several species are to be sown, divide tray into sections. Press seed gently into compost.

growing from seed, so details for their care are given here instead of under each individual entry.

Spring is a good time to sow, but two important points should be noted before you begin: seed must be fresh and pots or seed trays must be absolutely clean. A suitable seed compost contains 2 parts sterilized loam, 1 part fine peat, 1 part coarse sand and 1 part perlite. Seed can be sown in pots — choose 3in/7.5cm pots so that the compost does not dry out too quickly — or in a seed tray. Put a 0.5in/1.25cm layer of perlite or gravel at the bottom of the pot or tray and then fill with compost to within 0.5in/1.25cm of the rim. It helps to spread a thin layer of fine sand on top of the compost to give a more level surface and hence closer contact for the seeds. To moisten the compost, place the container in shallow, tepid water containing a fungicide to combat the risk of seeds 'damping-off'. Wait until the moisture has soaked through the compost, then take the tray from the water and let it drain.

Press large seeds into the surface of the compost (or top layer of sand) about 0.5/1.25cm apart: smaller seeds, preferably mixed with a little sand, are scattered as evenly as possible over the surface and gently pressed on the compost. If several species are sown in one tray, mark the divisions between them by placing small strips of cane across the compost. Label your seeds directly after sowing, or you will never be able to identify the little green shoots that emerge.

Cover the tray with a piece of glass and a sheet of newspaper and keep it in a temperature of 70–75°F/ 21–24°C, but not in the sun. Seeds of some species

Removing offsets

Offsets are removed in spring. Take the plant from its pot and break away compost from the roots. If these do not come away easily, cut with a sharp knife. Offsets with small or no roots should dry out for 2–3 days before potting-up. Bury the offset base about 0.5in/1.25cm deep in compost.

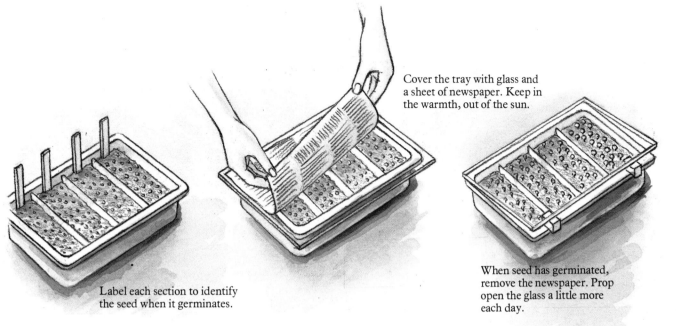

Cover the tray with glass and a sheet of newspaper. Keep in the warmth, out of the sun.

Label each section to identify the seed when it germinates.

When seed has germinated, remove the newspaper. Prop open the glass a little more each day.

Removing pads

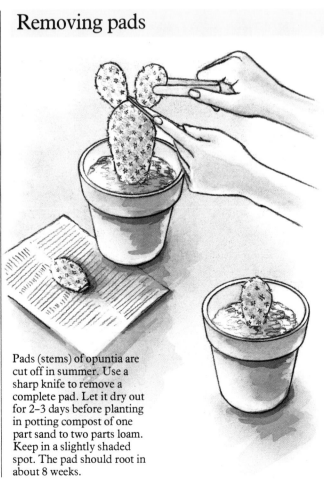

Pads (stems) of opuntia are cut off in summer. Use a sharp knife to remove a complete pad. Let it dry out for 2–3 days before planting in potting compost of one part sand to two parts loam. Keep in a slightly shaded spot. The pad should root in about 8 weeks.

may begin to germinate in a week, but others take two or three weeks. Wipe away any drops of moisture from the underside of the glass every day and water the tray from below, by putting it in a bowl of shallow, tepid water. The compost must stay moist, but not become sodden. As soon as the seedlings are visible, remove the paper and prop the glass open a little more each day until it can be removed.

If the seeds have been well spaced out, the seedlings — of desert cacti in particular — may thrive better if they are not transplanted until the following year. But if seedlings are at all crowded (with less than 0.5/1.25cm between them) they should be transplanted straight away. Move them to another tray, allowing 1in/2.5cm between each plant. Take the utmost care when moving these young seedlings for the roots are easily, and probably irremediably, damaged.

During winter months, keep the young plants at normal room temperatures. The following spring, move each one to a separate pot. A cactus will need only a 2in/5cm pot and will be content there for at least one year. Be in no great hurry to pot on.

Propagation of desert cacti by seed is particularly recommended for those which grow in the shape of columns or globes, are non-branching and do not grow offsets. While it is possible to propagate these by cuttings, to do so ruins the appearance of the plant and, in effect, sacrifices it. Examples of cacti better propagated from seed are astrophytum, cephalocereus, echinocactus, ferocactus and parodia.

Cacti and Succulents

OTHER WAYS WITH CACTI

The easiest cuttings to take are from offsets which appear at the base of the plant. These can be removed between late spring and late summer. Some offshoots are easily pulled away; others have to be cut. A clean cut is vital, and any offset which has had to be cut must be dried out (in a shady place) for two or three days. If the wound is not dried out before planting, the stem is likely to rot. Do not plant deeply; the base of the offset should be buried about 0.5in/1.25cm, enough to anchor it.

Cacti which grow from segmented stems (pads), such as opuntia, are also easy to propagate. Choose a pad which will least damage the appearance of the plant if it disappears. Cut it off completely and cleanly, dry the cut, and insert it just a little way into compost.

Some jungle cacti — schlumbergera and rhipsalidopsis, for example — have segmented stems. With these, take cuttings with two or three segments, dry them out and plant one segment deep.

Using cuttings from columnar or globular cacti, which neither branch nor produce offsets, is not recommended, unless the plant has grown too large or is far past its best. Cut at least 2in/5cm off the top of the column, dry it out and plant it. In time the bottom half of the parent plant will put out shoots and, when some 2in/5cm long, these can also be used as cuttings.

Globular cacti are more of a problem. The top half is hard to root and is best forgotten. The bottom half will put out shoots which can be used as cuttings.

SUCCULENTS WITHOUT SEEDS

As well as being grown from seed, succulents are propagated in a number of ways — by division, offsets, stem cuttings, leaf cuttings and rosette cuttings. These operations are explained in the plant entries. But there are common points to watch for.

Many succulents ooze large amounts of milky sap when cut. It is useless to insert the cutting in compost before this bleeding has stopped and the cut has hardened over, for a weeping cutting quickly rots. Some stop bleeding in a few hours; others take several days. To dry out a cutting, place it in a partly shaded spot with plenty of ventilation, so that the wound heals slowly.

Stem cuttings root more easily when they are ripe than do soft side shoots, and leaf cuttings in general take longer to root than stem cuttings. Leaf cuttings of crassulas and sedums usually give good results, however.

Offsets are no problem. They are produced round the base of a plant and are cut away from the parent, with a sharp knife, 0.5/1.25cm below the surface of the compost. Leave for a day to dry out before planting.

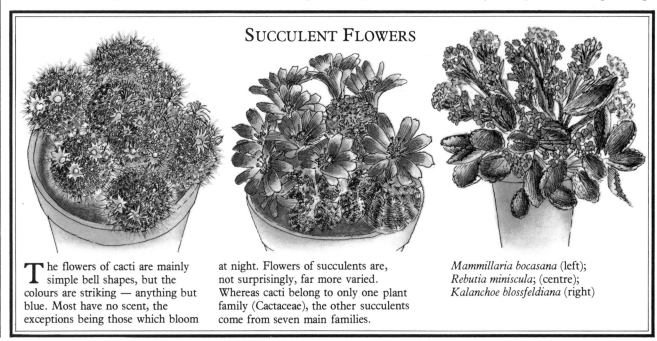

SUCCULENT FLOWERS

The flowers of cacti are mainly simple bell shapes, but the colours are striking — anything but blue. Most have no scent, the exceptions being those which bloom at night. Flowers of succulents are, not surprisingly, far more varied. Whereas cacti belong to only one plant family (Cactaceae), the other succulents come from seven main families.

Mammillaria bocasana (left); *Rebutia miniscula*; (centre); *Kalanchoe blossfeldiana* (right)

Aeonium

STEM CUTTINGS; SEED

Succulent which grows to 2ft/60cm. Temperatures required: summer, 65–75°F/18–24°C; winter, 45–50°F/7–10°C. *Aeonium haworthii*, Pin-wheel, has tall, woody branches bearing rosettes of fleshy, blue-green leaves edged with red. Pink-tinged yellow flowers bloom in late spring. *Aeonium tabulaeforme* has a rosette of close-packed leaves in an inverted saucer shape, up to 12in/30cm in diameter.

Stem cutting of *Aeonium haworthii*

Stem cuttings of *Aeonium haworthii* are taken in spring. Cut away a complete rosette with 2in/5cm of stem. Leave the stem to dry out for a day. Dip the cut end in hormone rooting powder and plant in a mixture of half loam potting compost and half sand. Place in a propagator at 65–70°F/18–21°C. Rooting takes about four weeks. When growing well, acclimatise the cutting to lower room temperatures over 2 weeks. Pot-on, using two parts loam potting compost to one

part sand, when roots fill the pot.
Seed of *Aeonium tabulaeforme* is sown according to instructions given in the introduction, page 134, and is the only way to raise new plants of this aeonium, which forms just a single rosette.

Aloe

OFFSETS; SEED

Succulent which grows 12in/30cm tall. Temperatures required: summer, 65–75°F/18–24°C; winter, 45–50°F/7–10°C. *Aloe variegata*, Partridge Breasted Aloe, has rosettes of triangular-shaped leaves banded with white. Pink tubular flowers appear in spring or summer.

Offsets are removed when repotting in spring. They will appear around the base of the parent plant, but wait until they reach 2–3in/5–7.5cm tall, with several leaves, before detaching them. Remove plant from its pot and gently break away compost from the offsets. Cut away only those which have developed roots and plant in a mixture of two thirds loam potting compost and one third sand. Keep out of direct sun until well established. Compost should be just moist.
Seed is sown according to instructions given in the introduction, page 134.

Aporocactus

CUTTINGS; SEED

Desert cactus which trails to 3ft/90cm. Temperatures required: summer, 65°F/18°C; winter, 50°F/10°C – minimum 45°F/7°C. *Aporocactus flagelliformis*, Rat's Tail Cactus, has green trailing stems covered with brown spines. Deep pink tubular flowers appear in spring and early summer.

Cuttings are taken in summer. Choose a stem-tip cutting about 6in/15cm long and allow it to dry out for 2 or 3 days. Insert to a depth of 1in/2.5cm in a pot of one part sand to two parts loam potting compost. Place in a propagator at 70°F/21°C. Cuttings should root in 6–8 weeks. Remove from the propagator when well established.

Seed is sown according to instructions given in the introduction, page 134.

Cuttings of *Aporocactus flagelliformis*

Astrophytum

SEED

Desert cactus which grows to a height and girth of 8in/20cm. Temperatures required: summer, 65–70°F/18–21°C; winter, 45°F/7°C. *Astrophytum myriostigma*, Bishop's Cap, is a globular cactus of 4 to 5 ribs covered with irregular white scales. Yellow, funnel-shaped flowers with red throats bloom throughout the summer months.

Seed is sown according to instructions given in the introduction, page 134.

Cereus

SEED; CUTTINGS

Desert cactus which grows to 6ft/1.8m but can be contained to 2–3ft/60–90cm. Temperatures required: summer, 65–75°F/18–24°C; winter, 45°F/7°C. *Cereus peruvianus* has bluish-green columns of 5 to 8 ribs, from which brown spines protrude.

Seed is sown according to instructions given in the introduction, page 134, and is the most sensible method of propagation in this case because cuttings involve the virtual sacrifice of the plant.

Cuttings are taken by slicing off the top 6in/15cm of the column with a sharp knife. Leave to dry out for 2 or 3 days and then plant to a depth of 1in/2.5cm in a pot of one part sand to two parts loam potting compost. Place in a propagator at 70°F/21°C until growing well. The base of the column will produce branches from the cut surface. When these grow to 3in/7.5cm they can be cut away and treated as cuttings.

Crassula

STEM CUTTINGS; LEAF CUTTINGS; SEED

Succulent which grows to 3ft/90cm tall. Temperatures required: summer, 65–75°F/18–24°C; winter, 45–50°F/ 7–10°C. *Crassula aborescens*, Silver Dollar Plant, has grey-green, fleshy, nearly round leaves with red edges, carried on woody stems. White to pink flowers appear in spring. *Crassula falcata*, Sickle Plant, has grey-green, fleshy, sickle-shaped leaves. Scarlet flowers bloom in summer.

Stem cuttings are taken in late spring from non-flowering shoots. Leave the 4in/10cm cuttings to dry out for 2 days and remove the bottom pair of leaves, if necessary. Plant in a mixture of half peat potting compost and half sand. Keep in the warmth out of direct sunlight. Rooting takes about 3–4 months. When roots fill the pot, pot-on using a mixture of one third sand to two thirds loam potting compost. *Crassula falcata* may throw out new shoots from the base of the

plant and these can be used for stem cuttings when tall enough. Cut into the stem at compost level.

Stem cutting of *Crassula aborescens*

Leaf cuttings are taken in late spring. Cut away a complete leaf and leave it to dry out for 2 days. Insert the cut end a little way into a pot of half peat potting compost and half sand. Thereafter treat in the same way as stem cuttings. Most crassulas can be propagated in this way, but stem cuttings are more successful.
Seed is sown according to instructions given in the introduction, page 134.

Echeveria

LEAF CUTTINGS; OFFSETS; LEAF ROSETTES WITH STEM ATTACHED; SEED

Succulent which grows 18in/45cm tall. Temperatures required: summer, 65–75°F/18–24°C; winter, minimum 45°F/7°C. *Echeveria derenbergii*, Painted Lady, has small rosettes of grey-green leaves tipped with red. Orange-red flowers appear in early summer. *Echeveria harmsii* has rosettes of lance-shaped leaves on short stems. *Echeveria gibbiflora*, reaching 18in/45cm, has a single stem carrying mauve-pink flowers.

Leaf cuttings are taken in late spring from echeverias not producing offsets or rosettes of leaves. Cut away a complete leaf with a sharp knife and leave it to dry out for 2 days. Insert the cut end in a pot of half peat potting compost and half sand. Keep in the warmth but out of direct sun. Rooting takes 4–6 weeks. A small plantlet will eventually grow from the base of the leaf. When it can be handled easily, repot in two thirds loam potting compost and one third sand.

Leaf rosette of *Echeveria derenbergii*

Offsets are removed when they reach about 1in/2.5cm across. Break away from the parent plant, trim the stem to 1in/2.5cm and remove the bottom leaves. Insert the stem into a pot of two thirds loam potting compost and one third sand. The leaves should not touch the compost mixture. Keep in the warmth but out of direct sun. Rooting takes about 4 weeks.
Leaf rosettes of *Echeveria harmsii* are removed in late spring. Take a complete rosette of leaves with stem attached and treat in the same way as offsets.
Seed is sown according to instructions given in the introduction, page 134.

Echinocactus

SEED

Desert cactus with globes which reach 6in/15cm across. Temperatures required: summer, 65–70°F/18–21°C; winter, 50°F/10°C – minimum 45°F/7°C. *Echinocactus grusonii*, Golden Ball, forms a many-ribbed globe from which golden-yellow spines grow. It is unlikely to flower indoors.

Seed is sown according to instructions given in the introduction, page 134.

Echinocereus

SEED; CUTTINGS

Desert cactus which grows 8in/20cm tall and 3in/7.5cm across. Temperatures required: summer, 65–75°F/18–24°C; winter, 45°F/7°C. *Echinocereus pectinatus*, Hedgehog Cactus, is columnar with up to 20 ribs from which emerge pink spines, turning white. It branches when about 6in/15cm tall. Bell-shaped pink flowers appear in summer.

Seed is sown according to instructions given in the introduction, page 134.
Cuttings are taken when branches have reached a height of 2in/5cm. Cut them away at compost level. Leave to dry out for 2 or 3 days and then insert to a depth of 0.5in/1.25cm in a pot of one part sand to two parts loam potting compost. Place in a propagator at 70°F/21°C. Rooting should take 6–8 weeks. Remove from the propagator when growing well.

Epiphyllum

CUTTINGS(STEM SECTIONS); SEED

Jungle cactus which grows to 3ft/90cm. Temperatures required: summer, 65–70°F/18–21°C; winter, 45–50°F/7–10°C. *Epiphyllum* 'Ackermanii', Orchid Cactus, has flattened stems from which tiny bristles protrude. Large red, cup-shaped flowers bloom in late spring. The hybrid 'Cooperi' has yellow flowers.

Cuttings are taken in summer. Cut away

Epiphyllum 'Ackermanii'

6in/15cm tall branches at compost level. Allow to dry out for a day or two and then insert several to a depth of 1in/2.5cm in a mixture of one part sand to two parts loam compost. Place in a propagator at 70°F/21°C. Cuttings should root in about 4 weeks. When growing well, remove from the propagator.
Seed is sown according to instructions given in the introduction, page 134.

Euphorbia

STEM CUTTINGS; SEED

Succulent which grows to 3ft/90cm. Temperatures required: summer, 65–75°F/18–24°C; winter, 55°F/13°C. *Euphorbia milii*, Crown of Thorns, has bright green, oval leaves carried on stout, spiny stems. Bright red or yellow bracts appear from spring to summer.

Stem cuttings are taken in late spring, using a piece which is 4in/10cm long. Milky sap can be stopped by dipping the cutting in water and spraying the cut stem on the main plant. Allow the cutting to dry out for 2 or 3 days before planting in a mixture of half peat potting compost and half sand. Keep out of direct sun, but in a warm spot. Rooting takes 8–10 weeks. When the pot is filled with roots, pot-on in a mixture of two thirds loam potting compost to one third sand.
Seed is sown according to instructions given in the introduction, page 134.

Faucaria

OFFSETS; SEEDS

Succulent with leaves growing to 2in/5cm long. Temperatures required: summer, 65–75°F/18–24°C; winter, 45°F/7°C. *Faucaria tigrina*, Tiger's Jaws, has rosettes of grey-green, pointed leaves with toothed edges arranged in opposite pairs. Yellow daisy-like flowers bloom in autumn.

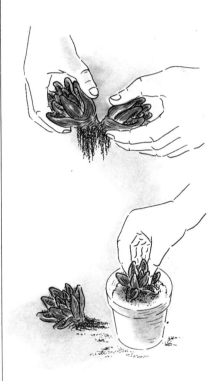

Offset of *Faucaria tigrina*

Offsets are removed when repotting in spring. Overcrowded clumps of roots can then be divided. Remove the plant from its pot and gently ease compost from the roots. Break away offsets, preferably with roots, and plant in a mixture of two thirds loam potting compost and one third sand. Any offsets removed without roots should be left to dry out for a day or two before potting-up. Keep in the warmth but out of direct sun until growing well. Do not allow the compost to become wet; offsets are likely to rot.
Seed is sown according to instructions given in the introduction, page 134.

Gasteria

OFFSETS; SEED

Succulent with leaves growing to 8in/20cm tall. Temperatures required: summer, 65–75°F/ 18–24°C; winter, 45°F/7°C. *Gasteria verrucosa*, Ox Tongue, has dark green, fleshy, pointed leaves covered with white hairs and arranged in opposite pairs one on top of the other. Red tubular flowers bloom from spring to early autumn. *Gasteria maculata* is similar, but larger, with strap-shaped leaves.

Offset of *Gasteria verrucosa*

Offsets are removed in late spring when repotting. Take the plant from its pot and gently ease away compost from the roots. Pull or cut away any offsets which have formed, selecting those with roots. Plant in a mixture of two parts loam potting compost and one part sand. Any offsets removed without roots should be left to dry out for a day or two before potting-up. Keep in the warmth but out of direct sun until well established.
Seed is sown according to instructions given in the introduction, page 134.

Haworthia

OFFSETS; SEED

Succulent with rosettes of leaves growing to 4in/10cm high and 6in/15cm across. Temperatures required: summer, 65–75°F/ 18–24°C; winter, 45°F/7°C. *Haworthia margaritifera*, Pearl Plant, has rosettes of dark green leaves covered with pearly-looking warts. Small, white flowers bloom in summer and early autumn.

Offsets are removed in spring from mature plants which have produced them. When repotting, remove the plant from its pot and break away compost from the roots, taking care not to damage them. Pull away offsets, preferably with roots, and plant in a mixture of two thirds loam potting compost and one third sand. Any offsets without roots should be left to dry out for a day or two before potting-up. Keep in the warmth but out of direct sun until growing well.
Seed is sown according to instructions given in the introduction, page 134.

Kalanchoe

STEM CUTTINGS; SEED

Succulent which grows to 18in/45cm. Temperatures required: summer, 60–65°F/16–18°C; winter, 55–60°F/13–16°C. *Kalanchoe blossfeldiana* has circular, fleshy, dark green leaves tinged red. Yellow, orange or red flowers on short stalks appear from early to late winter. *Kalanchoe tomentosa*, Pussy Ears, has rosettes of oval leaves, covered with white down and brown patches, carried on stems up to 18in/45cm tall. *Kalanchoe pumila* has pink-grey leaves with serrated edges covered with a whitish powder. Pink to violet flowers bloom in late winter.

Stem cuttings are taken in late spring, about 3in/7.5cm long. Should the plant be in flower, choose non-flowering stems. Leave the stems to dry out for 24 hours and then plant in a pot of half peat potting compost and half sand. Keep in the warmth in good light but out of direct sun. When growing well, with roots showing through holes in the pot base, pot-on using a mixture of two thirds loam

potting compost and one third sand.
Seed is sown according to instructions given in the introduction, page 134.

Lithops

DIVISION; SEED

Succulent which grows no more than 1in/2.5cm tall. Temperatures required: summer, 65–70°F/18–21°C; winter, 50°F/10°C. *Lithops lesliei*, Living Stones, has fleshy, flat-topped, grey-green leaves, spotted reddish-brown, just showing above the surface of the compost. Leaves grow in pairs, with yellow flowers emerging from the slit between them in late summer and autumn.

Division takes place in spring when repotting overcrowded groups of leaves. Remove the plant from its pot and gently ease away compost from the roots. Carefully pull the clumps of leaves apart, dividing into 2 pieces rather than detaching individual pairs of leaves. This causes less root disturbance and the divided plants will look more impressive. Plant each piece in a mixture of half loam potting compost and half sand. Keep in the warmth in good light, but out of direct sun, until recovered from the shock.
Seed is sown according to instructions given in the introduction, page 134.

Lobivia

OFFSETS; SEED

Desert cactus with globes growing to 4in/10cm in height and diameter. Temperatures required: summer, 65–75°F/18–24°C; winter, 45°F/7°C. *Lobivia hertrichiana*, Cob Cactus, has globes with up to 12 ribs from which brown spines protrude. Large scarlet flowers appear in summer.

Offsets are removed in spring. Take the plant from its pot and gently tease away compost from the roots. Select offsets which have developed roots and pull them away from the main plant; they should detach easily. Plant in a 3in/7.5cm pot of one part sand to two parts loam potting compost. Keep out of direct sunlight for a few weeks until well established.

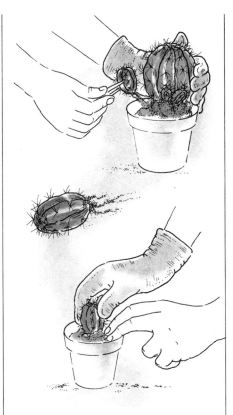

Offset of *Lobivia hertrichiana*

Seed is sown according to instructions given in the introduction, page 134.

Mammillaria

OFFSETS; SEED

Desert cactus with globes growing to 4in/10cm in height and diameter. Temperatures required: summer, 65–70°F/18–24°C; winter, 45°F/ 7°C. *Mammillaria bocasana*, Powder Puff, has clusters of blue-green globes about 2in/5cm across, covered with silky white hairs which hide hooked spines. Cream to yellow bell-shaped flowers bloom in spring. *Mammillaria elegans* is cylindrical, covered with short white spines. Red flowers emerge in spring on mature plants. *Mammillaria zeilmanniana* has 4in/10cm elongated globes covered with white and brown spines. Red flowers appear in summer.

Offsets are removed when repotting in spring. Take the plant from its pot and gently break away compost from the roots. As far as possible, choose offsets which have plenty of roots. Cut or break them away and plant in a 3in/7.5cm pot of one part sand to two parts loam potting compost. Offsets with little or no roots should be dried out for 2 or 3 days before potting-up. Keep in a slightly shaded spot until growing well.
Seed is sown according to instructions given in the introduction, page 134.

Opuntia

CUTTINGS OF SINGLE PADS; SEED

Desert cactus which grows to 12in/30cm tall, with a similar spread. Temperatures required: summer, 65–75°F/18–24°C; winter, 60°F/16°C, minimum 55°F/13°C. *Opuntia microdasys*, Prickly Pear, has flattened stems or pads covered with small yellow bristles. It produces yellow flowers, but this is unusual in indoor plants.

Cuttings are taken in summer. Cut off a complete pad using a sharp knife. Leave it to dry out for 2 or 3 days and then plant in a pot of one part sand to two parts loam potting compost. Keep out of direct sun in a slightly shaded spot. Rooting takes about 8 weeks.
Seed is sown according to instructions given in the introduction, page 134.

Rebutia

SEED; OFFSETS

Desert cactus with clumps of globes each about 2in/5cm across. Temperatures required: summer, 65–75°F/18–24°C; winter, 45°F/7°C. *Rebutia miniscula*, Red Crown, has clumps of small globes covered with short white spines. Flaring, trumpet-shaped flowers appear in late spring and summer.

Seed is sown according to instructions given in the introduction, page 134. This is a better method than using offsets.
Offsets are removed in summer. Take the plant from its pot and break away compost from the roots. Pull away one or two globes; they should detach easily. Plant in a mixture of one part sand to two parts loam potting compost. Keep in the warmth but out of direct sun until growing well. Rooting takes about 8 weeks.

MARK OF THE CACTUS

Cacti have a distinguishing mark which no other plants possess. This unique feature is called an areole. Areoles are plain to see on some cacti because they have large spines or hairs growing from them, but on other plants they are so small that they are easily missed. It is from areoles that the flowers grow, but each areole will produce only one flower in the life of the plant. Fortunately, as the cactus grows so do more areoles, so the plant can continue flowering.

Ferocactus latispinus, Devil's Tongue, is grown from seed, as described on page 134.

Rhipsalidopsis

CUTTINGS

Jungle cactus with stems growing to 12in/30cm long. Temperatures required: summer, 65–75°F/ 18–24°C; winter, 55°F/13°C. *Rhipsalidopsis gaertneri*, Easter Cactus, has stems of flattened segments which eventually arch over. Large, brick-red flowers emerge from the areoles on the last formed segments in spring.

Cuttings are taken in summer. Break off the end piece of a stem with 2 or 3 joints. Leave to dry out for a day or two and then plant to a depth of one joint in a mixture of one part sand to two parts loam potting compost. Keep in the warmth, out of direct sun, until growing well. Rooting takes 6–8 weeks.

Schlumbergera

CUTTINGS

Jungle cactus with stems reaching 12in/30cm long. Temperatures required: summer, 65–75°F/ 18–24°C; winter, 55–60°F/13–16°C. *Schlumbergera truncata*, Crab Cactus, has stems of notched segments. Pink to dark red flowers emerge from the last formed segment from early to midwinter.

Cuttings are taken in summer. Break off the end piece of a stem with 2 or 3 joints. Leave to dry out for 2–3 days and then plant to a depth of one segment in a mixture of one part sand to two parts loam potting compost. Keep warm but out of direct sun. Rooting takes 4–6 weeks.

Sedum

STEM CUTTINGS; SEED

Succulent with trailing stems to 3ft/90cm and upright plants to 12in/30cm. Temperatures required: summer, 65–75°F/18–24°C; winter, 45–50°F/7–10°C. *Sedum morganianum*, Burro's Tail, has trailing stems covered with small, pale green, overlapping leaves. Pink flowers may appear in summer. *Sedum sieboldii* 'Mediovariegatum' trails

Taking a cutting from *Sedum morganianum*

to 8in/20cm, its stems covered with red-edged, round, green leaves blotched yellow. *Sedum pachyphyllum*, Jelly Beans, has stems covered with pale green capsules tipped red.

Stem cuttings are taken in summer, 4in/10cm long. Remove the bottom leaves, exposing about 1in/2.5cm of stem, and leave to dry out for a day or two. Plant cuttings in a mixture of two thirds loam potting compost and one third sand. Bottom leaves should not touch the compost surface. Keep in the warmth, out of direct sun, until growing well.
Seed is sown according to instructions given in the introduction, page 134.

Senecio

STEM CUTTINGS; SEED

Succulent which trails to 3ft/90cm. Temperatures required: summer, 60–65°F/16–18°C; winter, 50°F/10°C.

Senecio rowleyanus, String of Beads, has stems covered with green globular leaves. White and purple flowers bloom in early autumn.

Stem cuttings are taken in summer, about 3in/7.5cm long. Make the cut just below a node. Remove the bottom leaf and insert several cuttings in a mixture of half peat potting compost and half sand. Keep in the warmth, but out of direct sun, until rooted. When the pot is full of roots, pot-on using two thirds loam potting compost to one part sand.
Seed is sown according to instructions given in the introduction, page 134.

Stapelia

DIVISION; STEM CUTTINGS; SEED

Succulent which grows to 6in/15cm. Temperatures required: summer, 65–75°F/18–24°C; winter, minimum 60°F/16°C. *Stapelia variegata*, Starfish Plant, has upright, fleshy, shiny, bright green finger stems. Yellow starfish-shaped flowers blotched with purple-brown bloom in summer.

Division takes place when repotting large clumps of stems in spring. Remove plant from its pot and gently break away compost from the roots. Pull or cut apart the stems, dividing into 2 pieces. Plant in a mixture of two parts loam potting compost to one part sand.
Stem cuttings are taken from individual stems with roots which can be cut away and planted in the recommended mixture. Cuttings with no roots should be left to dry out for 2 days before planting to a depth of 1in/2.5cm in a mixture of half sand and half peat potting compost. Keep in the warmth but out of direct sun. Pot-on when roots fill the pot, using the recommended mixture.
Seed is sown according to instructions given in the introduction, page 134.

Ferns

Most ferns are propagated easily by division of crowns, by division of rhizomes, or by removal of offsets. Growing new plants from spores is both more complicated and more fascinating, but requires far more time and patience on the part of the propagator. Those wanting a quick and straightforward increase of their fern stock should try the vegetative methods.

DIVISION

Dividing ferns with crowns is very simple for indoor varieties, which are usually small enough to be separated by hand. More stubborn plants may need some cutting with a knife, however. Great clumps of outdoor ferns are more difficult to part. Dig up the plant, then force two garden forks, back to back, into the middle of the clump and between – not through – the crowns. Use the fork handles as levers to prise the plant in two.

Dividing ferns with rhizomes is also easy but should be undertaken with care to avoid damaging the underground stem. Tease out the compost, or soil, to reveal the rhizome. Divide it in two or three pieces, rather than many, ensuring that each piece has several growing points. Then replant to the same — usually shallow — depth as before.

RAISING FERNS FROM SPORES

This method of propagation is not as straightforward. Some species are difficult and there is probably more chance of success with adiantums and pteris. Spores take the place of seeds in the life cycle of the fern. They develop on the back of the fronds and are almost microscopically small. Spores are enclosed in minute cases, called sporangia, which form larger groups called sori. These make the patterns to be seen on the back of the fronds.

As soon as the sori turn brown, remove the frond, put it in a clean paper bag and firmly tie the top. Keep in the warmth and in a week or so, the dust-like spores will have fallen off into the bag. Tap them out on to a piece of paper or into an envelope, away from all draughts which can blow the whole lot away. The spores are then sown immediately, because fresh seeds

Growing ferns from spores

Starting point for a new fern (top, left), the brown clusters of sori, made up of millions of spores. When the sori burst, the ripe spores are planted in a shallow pot, covered with glass and left standing in a shallow tray of water. If kept in shade, in the warmth, a green growth called prothalli should appear after several weeks. This is the second stage in the fern's life.

Tiny plants emerge in about 2 months (bottom, left) and small groups can then be transplanted into trays. Before long the fronds, the final stage, will be recognizable. When large enough to handle, move the ferns into small pots.

germinate better than ones which have been stored.

The compost must be sterile and the container scrupulously clean. A suitable compost comprises one part of loam seed compost and two parts of moss peat. A 5in/12.5cm shallow clay pot (the type known as a pan) makes a good container. Fill the pan with sieved compost and, as a precaution against disease, pour boiling water over it. When the compost has cooled, level the surface.

Put a small pinch of spores on the tip of a knife blade and tap them evenly over the surface of the compost. Do not cover them with compost. Put a piece of glass over the pan and keep in the warmth and shade. Indoor ferns will need temperatures of 65–70°F/18–21°C to germinate, but hardy ferns will manage with rather less heat. Keep the pan standing in shallow water in a tray, so that the compost is always moist. Remove the glass only to wipe condensation from the underside every day.

After four weeks or so, the surface of the compost will be covered with what looks like green slime. In fact, these are hundreds of tiny prothalli, which form the next stage in the life cycle of the fern. An ever-moist compost is now absolutely vital, otherwise fertilization will not take place among the prothalli, and there will be no new ferns.

The final stage begins with the emergence of tiny plants, about two months later. Using a small notched stick, prick out small lumps of the prothalli into a tray of compost identical to that used before, and plant them 1in/2.5cm apart, no deeper than they were before. After a week or so, begin to harden them off by lifting off the glass for a little longer each day. When young fronds are growing well, remove the glass altogether. Continue to water from below. When large enough to handle, separate each fern and transplant into small individual pots. Pot-on into a succession of slightly larger pots, before the original pots become full of roots. When mature, the plants will need less frequent repotting.

Outdoor ferns are moved into cold frames for hardening off before being planted in the garden.

FERN FRONDS

The attraction of ferns, which lack flowers, is mainly in the shapes of their fronds. They may be bold, as in the Bird's Nest Fern (*Asplenium nidus*, below left) and the huge Staghorn Fern (*Platycerium*, see page 147) or delicate, as in the Maidenhair Fern (*Adiantum*, below right). Textures also differ greatly: the Bird's Nest has a shiny smoothness; the Maidenhair a filmy softness; the Staghorn is covered in white velvety scurf.

Adiantum

DIVISION OF RHIZOMES; SPORES

Grows 12–18in/30–45cm high, with fronds to 3ft/90cm, depending on the species. Temperatures required: summer, 60–65°F/16–18°C; winter, 55°F/13°C, not below 50°F/10°C. *Adiantum capillus veneris*, Maidenhair Fern, is the smallest form, with lacy foliage on black wiry stems, followed by *Adiantum cuneatum* (syn, *A. roddianum*), the Delta Maidenhair, with slightly coarser, triangular leaflets. The largest is *Adiantum tenerum*, the Fan Maidenhair; the 'Farleyense' variety of this species has crimped and frilled leaflets.

Division of rhizomes takes place when repotting in early spring. This is the recommended method, as raising from spores is tricky (though less so than with other ferns). Break up the root ball into 2 or 3 equal pieces and plant in peat potting compost. New plants will not have the symmetry of the original fern and will take some time to grow into a rounded shape again. If the root is broken into small pieces, each with only a few fronds, it will be even longer before a well-shaped plant develops.

Spores are propagated according to instructions given in the introduction, page 143.

Asplenium

PLANTLETS ON LEAVES; SPORES

Fronds grow up to 2ft/60cm long. Temperatures required: summer, 65–70°F/18–21°C; winter, 65–70°F/18–21°C, but *Asplenium bulbiferum* will survive down to 50°F/10°C. *Asplenium bulbiferum*, Hen and Chicken Fern, has bright green foliage, rather like carrot tops. Small plantlets grow from bulbils on the surface of the pinnae, and are used for propagation. *Asplenium nidus*, Bird's Nest Fern, looks completely different, forming a shuttlecock-like rosette of bright green, glossy, lance-shaped fronds.

Plantlets are removed from *Asplenium bulbiferum* when they have grown 4 or more small fronds. Carefully detach from the parent and plant several in a pot of peat potting compost, or a mixture of loam, peat and sand. Press gently into the compost, water, and place in a propagator. Keep at 65–70°F/18–21°C. When new growth appears, remove from propagator.

Spores are propagated according to instructions given in the introduction, page 143. This method is usual for raising new plants of *Asplenium nidus*.

Blechnum

OFFSETS; SPORES

Fronds grow to 3ft/90cm long. Temperatures required: summer, 65–75°F/18–24°C; winter, 55°F/13°C, but nearer 60°F/16°C is preferable. *Blechnum gibbum* has a shuttlecock formation of light green, shiny fronds, deeply divided into tapering leaflets.

Offsets are removed in spring when repotting. They may be thrown up round the base of the parent plant. Take the plant from its pot and carefully tease away compost from the offset. Cut it away with a sharp knife, ensuring that there are plenty of roots to support it. If there are just a few, leave the offset for another year before detaching it from the parent plant. Plant in a peat potting compost or equal parts of peat, loam and sand.

Spores are propagated according to detailed instructions given in the introduction, page 143.

Ferns

Cyrtomium

DIVISION; SPORES

Fronds grow up to 18in/45cm long. Temperatures required: summer, 65°F/18°C; winter, 55–60°F/13–16°C, and not below 50°F/10°C. *Cyrtomium falcatum*, Holly Fern, has dark green fronds covered with pairs of glossy green pinnae. The variety 'Rochfordianum' is more compact, with larger pinnae.

Division takes place in spring, when repotting. Remove plant from its pot and gently break away compost to reveal the rhizome from which the fronds and roots grow. Break the rhizome into pieces, each with at least 4 fronds and adequate roots. Preferably break the rhizome into only 2 or 3 pieces, rather than many smaller bits which would take months to achieve bushy growth. Plant each piece in a pot of peat potting compost or a mixture of equal parts of loam, peat and sand. Keep in the warmth in a slightly shaded spot until the fern recovers from the shock.
Spores are propagated according to instructions given in the introduction, page 143.

Davallia

DIVISION OF RHIZOMES

Fronds grow to 18in/45cm long. Temperatures required: summer, 65–70°F/18–21°C; winter, minimum 55°F/13°C. *Davallia canariensis*, Deersfoot Fern, has feathery, mid-green fronds with many small, triangular-shaped pinnae.

Division of rhizomes takes place in spring. The rhizomes, covered with brown hairs, creep over the surface of the compost. Cut off pieces of rhizome about 4in/10cm long, with 2 or 3 fronds attached. Pin down into pots of half peat potting compost and half sand, with the rhizomes touching the compost surface. Place in a propagator at 70°F/21°C. Some 6 weeks later, the rhizome should have thrown out roots and new growth will be showing. Accustom the plant over 2 weeks to lower room temperatures. Pot-on about 4 months later, using peat potting compost.

Nephrolepis

DIVISION; ROOTED RUNNERS

Nephrolepis exaltata

Fronds grow up to 3ft/90cm long. Temperatures required: summer, 55–65°F/13–18°C; winter, 55–60°F/13–16°C. *Nephrolepis exaltata*, Boston Fern, has arching, bright green fronds, bearing opposite pairs of pinnae. The cultivar 'Rooseveltii' has pinnae with wavy edges and 'Whitmanii' has pale green lacy pinnae.

Division takes place in spring when repotting. Large plants can be split in half. Remove plant from its pot and break away compost from the roots. Pull the root ball apart into 2 pieces; it may be necessary to begin with a knife, cutting a little way down into the root ball from the surface of the compost. Plant each piece in peat potting compost. Keep in the warmth in a slightly shaded spot until the plant recovers from the disturbance.
Rooted runners of *Nephrolepis exaltata* are removed in spring when repotting. The plant sends out hairy runners which take root and grow small plantlets if they come in contact with the compost. Encourage them to do so by pinning down the runner on the compost surface. Sever any rooted runners from the parent plant in spring and remove each plantlet, along with the roots to support it. Plant in a pot of loam potting compost.

Pellaea

DIVISION OF RHIZOME; SPORES

Stems grow 12in/30cm long. Temperatures required: summer, 55–65°F/13–18°C; winter, 55°F/13°C. *Pellaea rotundifolia*, Button Fern, has stems of leathery, oval, dark green, glossy pinnae carried on opposite pairs.

Division of rhizome takes place in spring when potting-on. Remove the plant from its pot and gently break away compost from the roots until the rhizome is visible. Cut the rhizome into pieces, each with several stems and plenty of roots. It is better to cut the rhizome in 2 or 3 pieces than into many small ones. Plant each piece in peat potting compost. Keep warm in a slightly shaded spot.
Spores are propagated according to instructions given in the introduction, page 143.

Phyllitis

DIVISION; SPORES

Fronds grow to 12in/30cm long. Temperatures required: summer, 60–65°F/16–18°C; winter, minimum 50°F/10°C. *Phyllitis scolopendrium*, Hart's Tongue Fern, has strap-shaped, pointed, bright green fronds on stems.

Division takes place in spring when repotting, and only with lush growing plants. Remove the plant from its pot and gently break away compost from the roots and rhizome. Divide the rhizome into 2 or, at most, 3 pieces, using a sharp knife. Smaller pieces will take years to look full-growing. Plant in a peat potting compost or a mixture of half-loam potting compost and half leaf mould.
Spores are propagated according to instructions given in the introduction, page 143.

Platycerium

OFFSETS; SPORES

Fronds grow 2–5ft/60–150cm long and spread up to 5ft/1.5m. Temperatures required: summer, 60–70°F/16–21°C;

winter, 60–65°F/16–18°C. *Platycerium bifurcatum* (syn *Platycerium alcicorne*), Staghorn Fern, has circular and forked fronds rather like the horns of a stag. *Platycerium grande*, Regal Elkhorn Fern, is similar but with much larger fronds.

Platycerium bifurcatum

Offsets are removed in spring when potting-on. They are thrown out occasionally from the base of the fern. Remove the plant from its pot and gently break away compost from the roots. Make sure that the offset has plenty of roots; if there are only a few, leave on the parent plant for another year. If ready to be removed, pull away, or cut off with a sharp knife if necessary. Plant in a mixture of peat and sphagnum moss and keep in a shaded spot until well established. Division of plants is not recommended; it both spoils the appearance of the fronds and damages the fronds.
Spores are propagated according to instructions given in the introduction, page 143.

Polypodium

DIVISION OF RHIZOME; SPORES

Fronds grow up to 3ft/90cm long. Temperatures required: summer, 60–70°F/16–21°C; winter, minimum 50°F/10°C. *Polypodium aureum* has light green fronds comprising opposite pairs of deeply divided pinnae and one terminal pinna, carried on tall stalks. The variety 'Mandaianum' has blue-green fronds with crinkly edges.

Division of rhizome takes place in spring when repotting. Branching pieces of the furry rhizome, part of which rests on the surface of the compost, can be cut away with fronds still attached. Remove the plant from its pot and carefully tease away compost from the rhizome. Cut away branching pieces about 3in/7.5cm long which have a number of fronds. Plant in a mixture of half loam potting compost and half leaf mould. As the rhizome will have few roots, if any, peg it down on the compost surface with a wire loop. Place in a propagator at 65–70°F/18–21°C until the rhizome has developed roots and is growing well. Rooting takes about 6 weeks. Accustom the plant to lower room temperatures over 2 weeks.
Spores are propagated according to instructions given in the introduction, page 143.

Pteris

DIVISION; SPORES

Grows to 18in/45cm high. Temperatures required: summer, 60–65°F/16–18°C; winter, 55–60°F/13–16°C. *Pteris cretica* 'Albolineata', Ribbon Brake, has stems bearing up to 4 strap-shaped pinnae, with a cream to white streak along their length. The species *Pteris cretica* has plain mid-green pinnae. The variety 'Winsetii' is larger, with crested tips to the pinnae. *Pteris ensiformis* 'Victoriae' is similar to *Pteris cretica*, with pinnae streaked silvery white.

Division takes place in spring when potting-on. Remove the fern from its pot and gently break away compost from the roots. Pull the roots apart, splitting the plant into 2 or 3 pieces only. If divided into many small pieces it will be many months before any begin to look attractive again. Plant each piece in peat potting compost or in a mixture of loam, peat and sand. Keep in a shaded spot until the plants are growing well.
Spores are propagated according to instructions given in the introduction, page 143. The method is probably more successful with pteris than with most other ferns.

OUTDOOR FERNS

Many of the ferns grown as indoor plants have hardy relatives which will grow well outdoors, providing a dense green backdrop to flowering plants. Some, such as *Adiantum pedatum*, deserve prominence in their own right. All can be raised from spores, some spreading themselves readily in this way, but for quick results, division of the rhizome is the most convenient method. Detailed instructions on raising plants from spores and division of rhizomes are given on page 143–144. Those plants described below are some of the more impressive hardy, outdoor ferns.

Adiantum pedatum, Hardy Maidenhair Fern, 12in/30cm. Dies down at end of year.
Blechnum tabulare, 3ft/90cm;
Blechnum spicant, 15in/37.5cm, evergreen in all but coldest areas.
Dryopteris filix-mas, Male Fern, 3ft/90cm;

Dryopteris pseudomas, Golden-scaled Male Fern. Both die down in winter.
Gymnocarpium dryopteris, Oak Fern, 12in/30cm. Deciduous.
Matteuccia struthiopteris, Ostrich Feather or Shuttlecock Fern, 3ft/90cm. Dies down in winter.
Onoclea sensibilis, American Sensitive Fern, 18in/45cm. Turns brown at first frost.
Phegopteris connectilis, Beech Fern, 12in/30cm. Needs lime-free soil. Will grow in shade.
Phyllitis scolopendrium, Hart's Tongue Fern, 12in/30cm. Grow in shade preferably; also sun. Evergreen.
Polypodium vulgare, Common Polypody, 12in/30cm. Green through autumn and winter. Dies down when new fronds appear in late spring.
Polystichum setiferum, Soft Shield Fern, 2ft/60cm. Evergreen. Dies down when new fronds appear in spring.

Orchids

Orchids are fascinating plants to grow, partly because they differ so from other house-plants. There is nothing really daunting about them, however. They need understanding rather than coddling, but must have a strict regime of hygiene if they are to escape fungal and viral diseases. A sterile potting medium is essential.

Orchids respond to three reasonably simple forms of vegetative propagation, the main method being division of the rhizomes. For a few orchids, stem cuttings or removal of plantlets are other choices.

DIVISION

Division of rhizomes is the simplest propagation method and gives good results with well-established healthy orchids. Choose a plant with a minimum of six pseudobulbs, and preferably eight to ten, so that after division each section of the rhizome will have three to five pseudobulbs. A plant divided with fewer may take several years to flower.

Tackle the operation in two stages. Begin in winter, by deciding which plants are large enough to divide. Cut through the rhizome, which lies just under the surface of the compost, so that each part has three or more pseudobulbs. The division must be made with a sharp knife, previously sterilized, to avoid exposure to disease. Then leave the plant alone until repotting time in spring.

For the second stage, remove the plant from its pot and ease away the old compost from the roots. The cut rhizome can then be separated easily into two parts. Cut away any dead roots. Repot each section in a pot which is not too large, but which has room for the existing root system and another two years' growth of the rhizome. Potting-on may then be necessary.

PLANTING

Orchids with rhizomes which grow forwards (for example, cattleyas and odontoglossums) should be planted with the oldest pseudobulb against one side of

Division of rhizomes

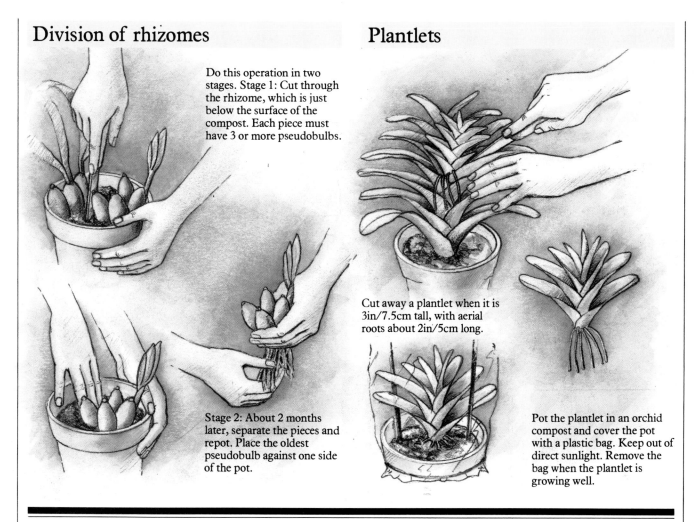

Do this operation in two stages. Stage 1: Cut through the rhizome, which is just below the surface of the compost. Each piece must have 3 or more pseudobulbs.

Stage 2: About 2 months later, separate the pieces and repot. Place the oldest pseudobulb against one side of the pot.

Plantlets

Cut away a plantlet when it is 3in/7.5cm tall, with aerial roots about 2in/5cm long.

Pot the plantlet in an orchid compost and cover the pot with a plastic bag. Keep out of direct sunlight. Remove the bag when the plantlet is growing well.

the pot, giving the rhizome space to grow towards the other side. Those which grow on all sides, as well as monopodial orchids, are planted in the middle of the pot. When planting, hold the orchid so that the base of growth is 1in/2.5cm below the rim of the pot. Then fill the space around it with compost. Among orchids which can be propagated by division are: cattleya; coelogyne; cymbidium; dendrobium; encyclia; epidendrum; laelia; miltonia; odontoglossum; oncidium; and paphiopedilum.

CUTTINGS AND PLANTLETS

Stem cuttings are useful for propagating mature, tall-growing orchids, such as the monopodial vandas and those epidendrums with cane-like stems. Such plants develop aerial roots, and cuttings are taken by removing the top section of stem directly below a set of roots. (For detailed instructions see the advice given for epidendrum and vanda in the following plant section.)

Plantlets may be found, with their own aerial roots, growing on dendrobiums, epidendrums and vandas. They can be carefully removed and planted when the roots are about 2in/5cm long, but it will be several years before they flower.

SEED

It is, of course, possible to grow orchids from seed. The old-fashioned way was to scatter the minute seed — always in danger of being blown away — on the surface of the live sphagnum moss in which an orchid of the same genus was growing. Orchids live in symbiotic association with various fungi, which must be present before the seed can germinate. Under this precarious system, it might be months before any sign of life appeared and during that time constant warmth and moisture were needed. Only the fittest germinated and survived.

Then science produced asymbiotic culture, in which chemicals replaced the orchid fungi. By this method, seeds are germinated in flasks containing a sterilized gell of agar and various chemicals. Germination is far more successful by this method, but it demands the strictest hygiene and a fair amount of equipment. Seedlings sold in these flasks have then to be transplanted into sterilized compost — a vulnerable stage in the system.

Now there is also meristem culture, in which plant cells, not seed, are used for propagation. So far, only orchids which produce pseudobulbs can be reproduced by this system, but it is hoped to extend the method to other varieties. Under sterile laboratory conditions, meristem cells are taken from pseudobulbs and made to proliferate. After various changes, they are removed to an agar gell containing nutrients in which young plants — all clones of the original parent — begin to grow. Apart from its commercial advantages, the merit of this propagation method is that the young plants are most likely to be virus-free. Sealed, small bottles, each containing a young orchid growing in agar jelly, can be bought from garden stores. Eventually, however, the bottle must be opened and the mericlone faces the outside world with its threat of fungi and viruses. This is the vulnerable point in the young plant's life.

ORCHID COMPOSTS

Orchid enthusiasts have their own preferred type of compost and will use no other if they can help it. Some of the older types are now scarce and extremely expensive, however, and perfectly acceptable compost can be made from sphagnum moss peat, perlite and coarse sand. Otherwise, a variety of ready-prepared orchid composts are on sale. Choose the correct variety; a terrestrial orchid needs a different type of compost from that required by an epiphyte.

Cattleya

DIVISION OF RHIZOME

Cattleya bowringiana

Pseudobulbs are 8in/20cm long and leaves grow to 5in/12.5cm. Temperatures required: summer, 65–75°F/18–24°C; winter, similar, *Cattleya bowringiana*, Cluster Cattleya, has 2 strap-shaped, leathery leaves growing from each pseudobulb. A long stalk of purple flowers, with a darker shaded lip, appears in mid-autumn.

Division of rhizome takes place in late winter. Cut through the rhizome, which lies just below the surface of the compost, and dust the cut ends with a fungicide. Each piece of rhizome should have at least 4–5 pseudobulbs. Do nothing more until late spring. Then remove the plant from its pot and carefully disentangle the roots of each section. Repot in a mixture of sphagnum moss peat, perlite and coarse sand or a ready-prepared orchid mixture. Each section may require a cane support until the roots anchor themselves in the mixture.

Coelogyne

DIVISION OF RHIZOME

Pseudobulbs grow 2in/5cm tall and leaves reach 12in/30cm. Temperatures required: summer, 65–75°F/18–24°C; winter, 45–55°F/7–13°C. *Coelogyne cristata* has 2 strap-shaped leaves on each round to egg-shaped pseudobulb. White flowers with yellow marked lips appear from winter to spring.

Division of rhizome takes place when flowering has finished. Choose a section with the most recently produced pseudobulb and at least three older ones. Cut through the rhizome, which lies just below the compost surface, and dust the cut ends with a fungicide. Now leave the plant undisturbed in the pot for 2 months. Then remove from its pot and carefully disentangle the roots of each piece of rhizome, trimming any roots which are obviously dead. Repot in a mixture of sphagnum moss peat, perlite and coarse sand or a ready-prepared orchid mixture. Support each piece with a cane until roots have a firm hold.

Cymbidium

DIVISION OF RHIZOME

Leaves grow to 20in/50cm and the flower stalk to 15in/37.5cm. Temperatures required: summer, 60–65°F/16–18°C; winter, 50–55°F/10–13°C. *Cymbidium* 'Peter Pan' is a hybrid with bright green,

narrow, sword-shaped leaves. Stems of greenish-yellow flowers with deep crimson lips appear from late spring to early autumn.

Division of rhizome takes place when flowers have died down. Remove the plant from its pot and ease away compost from the roots. Cut through the rhizome, ensuring that each piece has 3–4 pseudobulbs. Gently unravel the roots and trim any dead ones. Dust the cut rhizome ends with fungicide and plant each piece in a mixture of sphagnum moss peat, perlite and coarse sand, or a ready-prepared orchid mixture.

Dendrobium

DIVISION OF RHIZOME; PLANTLETS

Grows to 3ft/90cm. Temperatures required: summer, 70–75°F/21–24°C;

winter, 50–55°F/10–13°C. *Dendrobium nobile* has cane-like stems bearing small, narrow, strap-shaped leaves at the top. After the leaves fall in late winter, pink to purple flowers are produced, with yellow tips and a maroon blotch in the centre.

Division of rhizome takes place after flowering. Cut through the rhizome, found just below the surface of the compost; each piece should have 4 or more pseudobulbs, some of which have not produced leaves. These will be the potential flowering pseudobulbs. Any dead pseudobulbs should be removed completely. Dust cut ends with fungicide. Now leave the plant undisturbed for 2 months. Then remove it from the pot and disentangle the roots. Plant each piece of rhizome in a mixture of sphagnum moss peat, perlite and coarse sand, or a ready-prepared orchid mixture. Support with a cane if necessary.
Plantlets are removed when they reach about 3in/7.5cm tall, with a few leaves and roots. They may be produced at the base of stems, or near the top with aerial roots. Cut them away from the stem and plant in an orchid mixture. Cover with a plastic bag and keep out of the sun. When growing well, remove the bag.

Epidendrum

DIVISION OF RHIZOME; STEM CUTTINGS; PLANTLETS

Grows from 12in/30cm to 5ft/1.5m, depending on the species. Temperatures required: summer, 65–70°F/18–21°C; winter, 60–65°F/16–18°C. *Epidendrum radicans* (syn. *E. ibaguense*) has reed-like stems up to 5ft/1.5m tall with pale green, oval leaves. A flower spike of orange, red or pink flowers with fringed lips appears from late summer to autumn. *Epidendrum vitellinum* has small, egg-shaped pseudobulbs, each carrying 2–3 blue-green leaves. Orange-red flowers with a yellow lip bloom in autumn.

Division of rhizome takes place in late winter. All epidendrums can be divided if the pot is overcrowded with pseudobulbs. Cut the rhizome, found just below the surface of the compost, into pieces with at least 4 pseudobulbs. Dust the cut ends

with fungicide and leave for 2 months. Then remove the plant from its pot, break away compost from the roots and gently disentangle them. Pot each piece of rhizome in a mixture of sphagnum moss peat, perlite and coarse sand or a ready-prepared orchid mixture.

Epidendrum radicans

Stem cuttings are taken in spring. Epidendrums producing tall, cane-like pseudobulbs can be reduced in size by taking cuttings from the top of the pseudobulb. Aerial roots are produced on stems and the cut should be made just below a clump of roots. Cuttings should be at least 10in/25cm tall, with plenty of leaves. Moisten the aerial roots to make them more pliable and carefully arrange them in the pot, filling it with orchid compost. The cutting may require a cane support. New growth will appear from the cut stem of the original plant.
Plantlets of cane-like pseudobulbs are removed when they reach about 3in/7.5cm high, with a few leaves. The plantlets have aerial roots growing from the point where leaves join the stem (node). When these roots are about 2in/5cm long, cut the plantlet away from the stem, with the aerial roots, and plant in an orchid mixture. Cover with a plastic bag and keep out of direct sunlight. When the plantlet is growing well, remove the plastic bag.

Orchids

Laelia

DIVISION OF RHIZOME

Pseudobulbs grow to 4in/10cm tall and leaves to 12in/30cm. Temperatures required; summer, 60–65°F/ 16–18°C; winter, 50°F/10°C. *Laelia anceps* has 1–2 strap-shaped leaves growing from each pseudobulb. Pink to purple flowers, with a darker lip blotched yellow at the base, bloom from autumn to early winter.

Division of rhizome takes place in early spring, after flowering. Cut the rhizome, found just below the compost surface, into pieces with at least 4 pseudobulbs (the most recent and the next 3). Dust the cut ends of the rhizome with fungicide. Do nothing further for 2 months. Then remove the plant from its pot, break away compost from the roots and carefully disentangle them. Remove any dead roots and dark brown, leafless pseudobulbs. Pot each piece of rhizome in a mixture of sphagnum moss peat, perlite and coarse sand or a ready-prepared orchid mixture. Provide a cane support.

Lycaste

DIVISION OF RHIZOME

Pseudobulbs reach 4in/10cm long and leaves grow up to 12in/30cm. Temperatures required: summer, 65–70°F/18–21°C; winter, minimum 55–60°F/13–16°C. *Lycaste deppei* usually produces pairs of leaves from the flattened pseudobulbs. Flowers with pa'e green sepals, spotted reddish-brown, and white petals with yellow lips, spotted red, appear in spring, carried singly on 6in/15cm stems.

Division of rhizomes takes place in spring when flowering is over. Cut the rhizome, which grows just below the compost surface. Divide into pieces with at least 3 pseudobulbs. Dust the cut ends with fungicide and leave undisturbed for 2 months. Then remove the plant from its pot and carefully disentangle the roots. Plant each piece in a mixture of sphagnum peat moss, perlite and coarse sand or a ready-prepared orchid mixture.

Miltonia

DIVISION OF RHIZOME

Grows to 15in/37.5cm with a 24in/60cm flower stalk. Temperatures required: summer, 65–70°F/18–21°C; winter, 60–65°F/ 16–18°C. *Miltonia vexillaria*, Pansy Orchid, has strap-shaped leaves which grow from small pseudobulbs. Pink to red pansy-shaped flowers with yellow markings appear from spring to early summer.

Division of rhizome takes place when flowering has finished. Select plants with large clumps of pseudobulbs. Remove the plant from its pot and cut through the rhizome. Divide into pieces with at least 3–4 pseudobulbs and carefully disentangle the roots. Dust the cut ends with fungicide before planting each piece in a mixture of sphagnum moss peat, perlite and coarse sand, or a ready-prepared orchid mixture. Keep in a slightly shaded spot out of direct sun.

Odontoglossum

DIVISION OF RHIZOME

Pseudobulbs grow to 4in/10cm and leaves to 10in/25cm. Temperatures required: summer, 60–70°F/16–21°C; winter, 60°F/16°C. *Odontoglossum grande*, Tiger Orchid, has 2 strap-shaped leaves growing from each pseudobulb. Bright yellow flowers, striped reddish, appear in autumn.

Division of rhizome takes place in spring. Remove the plant from its pot and cut the rhizome to give pieces with at least 4 pseudobulbs. Break away compost from the roots and carefully disentangle them. Dust the cut ends of the rhizome with a fungicide. Plant each piece of rhizome in a mixture of sphagnum moss peat, perlite and coarse sand, or a ready-prepared orchid mixture. Keep in a slightly shaded spot out of direct sunlight.

Oncidium

DIVISION OF RHIZOME

Grows to 10in/25cm with a 2ft/60cm flower stalk. Temperatures required:

summer, 65–70°F/18–21°C; winter, 50–55°F/10–13°C. *Oncidium ornithorhyncham* has small pseudobulbs bearing 2 strap-shaped leaves. Rose pink flowers with a yellow crested lip appear from autumn to winter.

Division of rhizome takes place in spring. Remove the plant from its pot and ease compost away from the roots. Divide and cut the rhizome into 2 pieces, each with at least 4 pseudobulbs. Carefully disentangle the roots to each piece of rhizome and dust the cut ends with a fungicide. Plant in a mixture of sphagnum moss peat, perlite and coarse sand, or a ready-prepared orchid mixture. Keep the plants out of direct sunlight.

Paphiopedilum

DIVISION OF RHIZOME

Leaves grow to 12in/30cm and flower stalk to 18in/45cm. Temperatures required: summer, 60–65°F/ 16–18°C; winter, similar. *Paphiopedilum callosum*, Slipper Orchid, has dark green, strap-shaped leaves mottled light green. White flowers with maroon and green stripes, and light purple lips, bloom from late winter to spring.

Division of rhizome Unlike most orchids, paphiopedilums have no pseudobulbs. The leaves grow directly from the rhizome, which is found just below the surface of the compost. Remove the plant from its pot and break away compost from the roots. The rhizome should be divided into pieces with 3 or 4 clumps of leaves. It may be possible to pull apart the rhizome; otherwise cut it with a knife. Gently disentangle the roots and dust the cut ends of the rhizome with a fungicide. Plant each piece in a mixture of sphagnum moss peat, perlite and coarse sand, or a ready-prepared orchid mixture. Keep in a slightly shaded spot away from direct sunlight until growing well.

Phalaenopsis

PLANTLETS

Leaves grow up to 18in/45cm long and the flower spike to 2ft/60cm.

·Temperatures required: summer, 65–70°F/18–21°C; winter, the same. *Phalaenopsis* hybrids, Moth Orchids, produce a short, stout stem about 3in/7.5cm long from which alternating strap-shaped leaves emerge. Aerial roots grow from the stem. Moth-shaped pure white to pink flowers, sometimes spotted red, bloom on long arching stems.

Plantlets are removed when they have several leaves and are about 3in/7.5cm tall. These new plants will sometimes grow from the end of old flower spikes and from the base of the plant. Leave them to develop until the roots are about 2in/5cm long, then cut them away from the parent and plant in a mixture of sphagnum peat moss, perlite and coarse sand, or a ready-prepared orchid mixture. Cover with a plastic bag until growing well and keep in a warm place.

Vanda

STEM CUTTINGS; PLANTLETS

Grows to 12in/30cm. Temperatures required: summer, 65–70°F/ 18–21°C; winter, 50–55°F/10–13°C. *Vanda cristata* has opposite pairs of layered, strap-shaped leaves. Aerial roots grow from the stem. Yellow-green flowers bloom in early summer.

Stem cuttings are taken when a mature plant has produced aerial roots. After flowering has finished, cut a piece of stem, 6in/15cm long, just below a group of aerial roots. Soak the roots in water to make them more pliable and carefully arrange in a pot, trying not to break them. Fill the pot with a mixture of sphagnum moss peat, perlite and coarse sand, or a ready-prepared orchid mixture. Support the cutting with a cane if necessary. The cut stem of the original plant should put out new growth.

Plantlets are removed when they have several leaves and are 3in/7.5cm tall. Plantlets may be produced from nodes on the stem, or from the end of a flower stalk. When they have roots about 2in/5cm long, cut them away from the parent and plant in orchid mixture. Cover with a plastic bag until growing well and keep out of direct sun.

HOW ORCHIDS GROW

M ost orchids grown as houseplants are epiphytes, or air plants. In their natural habitat they have no roots in the soil but live above ground, on another plant, probably a tree. Terrestrial orchids — the paphio-pedilums, for example — are those which grow in the ground.

Orchids have two forms of growth. Monopodial orchids (such as vandas) have a single stem growing vertically. Sympodial (many-stemmed) orchids, such as cattleyas, have a horizontal rhizome. Each year, another swollen upright stem — a pseudobulb — grows from the rhizome. It stores food and water and looks like a bulb, but is not. From the pseudobulb grow one or two leaves, a flower sheath, and then a flower. After flowering, the orchid becomes dormant. When growth starts again, the past year's pseudobulb produces an eye. This becomes a shoot, which turns to grow upwards and swells to form a new pseudobulb — and so on each year.

Horizontal rhizome of *Cattleya* (above), a sympodial (many-stemmed) orchid. The single vertical stem of *Vanda* (below) denotes a monopodial orchid.

Plants from Pips

Pips are, of course, nothing more nor less than seeds. Usually they are discarded when we eat the fruits that contained them. But instead of throwing them all away, a few can be sown, just for fun, and soon our living rooms could be sprouting such exotica as avocados, lychees, mangoes, coffee and dates. Few will produce the fruits they were taken from and so have been given the title 'Fun Plants'.

Although great pleasure can be gained from growing them, some 'fun plants' are more satisfying to raise than others. An avocado pit can produce a very graceful shrub, as long as the growing tip is removed when the plant is still small, to make it grow bushier. If you hesitate to do this because it seems brutal, the end result will be a mophead of leaves on a spindly stem.

Orange and lemon pips give a good reward. Many of the smaller types of oranges will, in time, produce blossom and fruit, but there is no guarantee that the seed will breed true. Lemons will also grow attractive-looking fruits, which stay on the tree a long time while ripening.

Most pip growers remain content to propagate perennials; the most popular are given in the following plant entries. But some of the annual fruits and vegetable fruits we eat produce so many seeds that it is almost a challenge to sow them. Nothing but a little time is lost if the results do not fulfil expectations. A melon, for instance, is profligate in the production of seeds and, if sown straight away, they will germinate readily. A few scooped from a melon in summer, washed and dried and planted in peat seed compost, will soon provide decorative trailing plants, although they will not have time to fruit before the plant dies down in winter. Of the vegetable fruits the sweet pepper is an obvious choice. Closely sown seed makes an attractive grouping of young plants.

Not all seeds will germinate when sown straight away; some grow only after a period of dormancy. But it costs nothing to experiment.

Pips normally disregarded as rubbish can produce attractive and interesting houseplants.

Avocado

SEED

Persea gratissima. Tender evergreen, which indoors can grow to 6ft/1.8m. Its elliptical leaves grow to 12in/30cm long. Virtually a do-it-yourself plant, since they are rarely seen on sale.

Seed comes from the 'pit' (stone) of a ripe avocado. Soak in tepid water for 48 hours, keeping the water warmish by standing the bowl on a radiator. The seed can be rooted in water, with the base just touching the surface of the liquid, but it is far simpler to grow in compost from the start. Use a 4in/10cm pot and a peat potting compost. Plant with the fat end downwards and the pointed end just showing above the surface of the compost. Keep in the light. Provide a temperature of 70°F/21°C and keep the compost moist but not sodden. Germination may take anything from 10 days to 5 weeks; be patient, but if no shoot appears long after that you can assume that the pit was infertile. If a shoot does appear, wait until it reaches 6–8in/15–20cm and then cut it back to 4–5in/10–12.5cm. Only this drastic action can make the plant grow bushy instead of long and lanky. After a couple of weeks or so the dormant buds will show signs of life. After a few months, transplant to a 6in/15cm pot. If it flourishes it will eventually need a 10–12in/25–30cm pot or even a small tub.

Citrus Fruits

SEED; CUTTINGS

Citrus sinensis, Sweet Orange; *Citrus aurantium*, Seville Orange, is sour but blooms profusely, has a lot of pips and is quite hardy. *Citrus reticulata* includes the very pippy Tangerine; the Satsuma (frequently pipless); and the Clementine, which has pips if the blossom was fertilized but any potential fruit would not be true to the parent plant. The pips of these small oranges produce attractive looking plants which are hardy, mature quickly, blossom profusely, and are capable of producing fruit even when restricted to 12in/30cm pots. (But do not depend on that.)

Growing *Persea gratissima* from an avocado stone. Cut back the shoot for bushy growth.

Seed is sown towards the end of winter or early spring. Use fresh pips and sow 0.5in/1.25cm deep, preferably in peat pots so that the roots are not disturbed when transplanting. Temperatures

between 60–70°F/16–21°C are suitable. Transplant into potting compost in a 4in/10cm pot. When the plant is 12in/30cm high, pinch out the growing tip to encourage bushiness.

Citrus grandis, Grapefruit. Treat the pips in the same way as orange pips. The plant grows larger, with perfumed blossom.

Citrus limonia, Lemon. Notable among citrus fruits for its hardiness, but must not be exposed to frost. Blossom appears at various times during the year and the fruits stay on the tree to ripen the following year. Indoors it should reach full bearing when around 7ft/2.10m tall. The seeds do not breed true; the alternative is to take cuttings.
Seed is sown at any time of the year, in temperatures above 55°F/13°C. Sow one pip to a small pot of peat seed compost. Growth is rapid, so transplant to a 4in/10cm pot of potting compost.
Cuttings are taken in summer. Use semi-ripe cuttings, 4in/10cm long, and insert each one in a small pot of potting compost with added coarse sand. They root in a month or so.

Citrus aurantifolia, Lime. A bushier tree and less hardy than the lemon. Grow from pips, as for oranges.

Coffee

SEED

Coffea arabica. Grows indoors to 4ft/1.2m, with glossy dark green leaves, but rarely produces the fragrant white flowers or red berries. Buy the seed (the so-called beans) from a seed merchant and not (roasted) from a coffee shop.

Seed is sown in early spring or spring. Plant 0.5in/1.25cm deep in peat seed compost, one to each small pot. A temperature of 80°F/27°C, or very near, is essential. Germination takes 3–4 weeks. Transplant seedlings when 3–4in/7.5–10cm high. The following spring, transfer to 5in/12.5cm pots, using a loam potting compost. Beyond the seedling stage they need a temperature of 65–75°F/18–24°C in summer, and no lower than 60°F/16°C in winter.

Plants from Pips

Date

SEED

Phoenix dactylifera, Date Palm, has graceful bluish glaucous fronds; fairly easy to grow if you have patience.

Raising *Phoenix dactylifera* from seed. Early leaves (top) are unlike the fronds of the mature palm (bottom).

Seed comes from the date stone. Scrape along its length with sandpaper so that moisture from the compost can penetrate more easily. Plant in peat potting compost and keep in a propagator at 75°F/24°C to encourage early germination; in cooler temperatures the stones may not germinate for almost 3 months. Early leaves are not frond-like, but have smooth edges, and the first palm-like leaves take a couple of years to appear. Dates there will never be, nor will the palm ever grow to the 120ft/36m that it reaches in nature.

Lychee

SEED

Litchi chinensis has glossy, lance-shaped leaves. Grown indoors it reaches about 3ft/90cm, but will never have flowers or the plum-sized fruits with a crimson warty rind.

Seed comes from the glossy brown stone, which is planted in a 3in/7.5cm pot of peat potting compost. It will germinate at a temperature of 65°F/18°C, but often slowly. In winter the minimum temperature should be 60°F/16°C. Move to a larger pot the following spring.

Mango

SEED

Magnifera indica, tropical plant with narrow, dark green leaves. As a houseplant it grows into an attractive little tree. Inside the orange coloured flesh of the fruit is a large stone.

Seed comes from the stone. Plant in a 4in/10cm pot of peat potting compost and germinate in a temperature of 70–75°F/21–24°C. When it has grown to 10in/25cm, pinch out the growing tip to make the plant bushier. Pot-on, when roots fill the pot, into a 6in/15cm pot. The minimum winter temperature is 60°F/16°C.

Peach

SEED

Prunus persica makes a dainty houseplant when young, and can later be planted out in the garden in the hope that it may fruit.

Seed comes from the stones of very ripe peaches. Soak in water for 2 days. Slightly crack the stone with utmost care so that the kernel is not damaged. This will let in a little moisture. Sow 1in/2.5cm deep, using a peat pot so that transplanting does not disturb the roots, which peaches cannot tolerate. Cover several pots with a polythene bag to keep in moisture. Germination may take weeks or months. During this time keep in a warm room

and as soon as a shoot appears, remove the pots from the bag. Transplant each seedling, in its peat pot, into a 3in/7.5cm pot, packing potting compost round it without disturbing the roots. A little later move to 5in/12.5cm pots. If, in time, they are moved outdoors, do this when the weather has warmed up in spring and plant against a sunny wall where they are to grow. Whether they will fruit to expectation is another matter.

Plum

SEED

Prunus domestica is attractive as a foliage houseplant when young, but this soon diminishes. It is not worthwhile trying to grow it on outdoors.

Seed is sown in late winter. Plum stones have to be overwintered outdoors beforehand in a box of peat to induce them to germinate. Crack the stone slightly if it is still hard, as recommended for peaches. Thereafter also follow the same routine as for peaches.

Pomegranate

SEED

Punica granatum, tropical or sub-tropical tree or shrub with glossy green leaves on red stems. Since it is deciduous, its attraction is limited to the summer months. It may reach 6ft/1.8m indoors unless pruned every year. It will grow outdoors in the warmer areas of temperate regions, especially if given the shelter of a south-facing wall.

Seed is sown in 3in/7.5cm pots of peat seed compost, several to each pot. Germinate at 60–65°F/16–18°C. Transplant the sturdiest seedlings to single pots. After the leaves fall in the autumn, keep in a frost-free room until growth begins again in early spring.

Sprouting Seeds

Nothing in gardening brings faster results than sprouting seeds; three days or so from sowing to eating should tax nobody's patience. Not that you can sow and forget, for the snag about sprouting seeds is that they have to be rinsed every day, or they will grow sour.

Most seeds are sprouted in jars. Inevitably, with their growing popularity, special sprouters have been designed, largely to make the chore of rinsing easier, but many jars are suitable. A wide jar gives the seeds a better chance to germinate. A tray can be used instead of a jar, however, and the seeds are then sprouted on a damp base, in the way that mustard and cress can be grown on blotting paper.

When using a jar, first soak the seeds overnight in cold water; they begin to swell, which shortens sprouting time. The next morning, rinse and drain them. Put a layer of seeds in the jar, not more than 0.75in/2cm deep. Cover the top of the jar with a piece of muslin held firmly by an elastic band. Place the jar on its side on a plate or in a shallow container. Keep the bottom end of the jar raised a little with a piece of wood, so that excess water can drain from the seeds; never leave them standing in water. Put the jar in a warm place, in a temperature of 55–70°F/13–21°C. If the seeds are germinated and grow in the dark (an airing cupboard is usually suggested) the shoots will be white and crisp. Those germinated in the dark and then brought into the light will be greener and softer. Whether in light or darkness, the seeds must be kept warm and moist, but not in a stuffy atmosphere or they will turn mouldy. Rinsing is vital. Morning and evening is enough for some seeds; others need more (see individual entries for requirements). Turn the seeds into a fine kitchen sieve and wash (not too fiercely) with cold water under a tap. Then drain them and return to the jar. Otherwise, remove the muslin cover and half fill the jar with water. Then replace the

---

Seeds can be sprouted in jars, on dishes, or, like the mung beans and alfalfa here, in tiered containers manufactured for the task.

Sprouting Seeds

cover securely and pour out the water through it.

When using a tray, cover the base with several layers of damp kitchen paper towelling. Spread on top of it seeds which have been soaked overnight, rinsed and drained. Cover with another layer of paper, but remove this when the seeds begin to germinate. The tray can be enclosed in a plastic bag so that the seeds do not dry out rapidly; on the other hand they should never be left standing in water after rinsing. The tray must be kept in the warmth, and in the dark if white shoots are needed. To grow green shoots, move the tray into the light when the seeds have germinated, but not into sunlight. Seeds grown in trays have to be rinsed as frequently as those in jars. This is done by running water over them, which can be tricky.

Mung beans and lentils are candidates for tray treatment, but are liable to root into the moist base and may have to be cut, not pulled, away. Sprouts need close watching since they soon grow past their best. Sprout as few as you need at a time and as often as you wish.

Seeds for sprouting can be grown in a shallow plastic tray between layers of kitchen paper, or in a muslin-covered jar. They must be rinsed regularly and drained thoroughly.

Alfalfa

SEED

Medicago sativa has a fresh garden pea flavour. Sprouting takes 3–5 days, in the light for greening. Rinse twice a day. Harvest when shoots are 1–2in/2.5–5cm long.

Beans

SEED

Phaseolus angularis, Adzuki bean. Produces sweet, nutty, crisp shoots. Sprouting takes 3–6 days in darkness on a tray or in a container. Rinse 4 times a day. Harvest when 0.5–1in/1.2–2.5cm long.

Phaseolus aureus, Mung bean. Produces white, crisp shoots. Sprouting takes 3–8 days in the dark. Rinse 3 times a day. Harvest when 0.5–3in/1.2–7.5cm long.

Chick Pea

SEED

Cicer arietinum, the chief pulse crop of India, rich in protein; as sprouts often eaten with curry. Sprouting takes 3–4 days. Rinse 4–5 times daily. Harvest when 0.5in/1.2cm long.

Fenugreek

SEED

Trigonella foenumgraecum. A typical ingredient of curries, crisp when raw.

HEALTH WARNING

Make absolutely certain that the seeds you buy for sprouting have *not* been treated with chemicals; seeds meant for sowing in the garden may have been dressed with a fungicide or insecticide. A good health food shop is often the most reliable place to buy viable seeds for sprouting. Ask to be certain.

Sprouting takes 3–4 days, in the light for greening. Rinse twice a day. Harvest when 0.5in/1.2cm long for strong curry flavour, or leave 3–4 more days when shoots are 3in/7.5cm long for a milder flavour.

Grains

SEED

Several grain sprouts add a new flavour and texture to soups and salads. Among these are buckwheat. Sprouting takes 2–4 days; rinse once daily, harvest at 0.5in/1.2cm. Triticale (a hybrid of wheat, *Triticum*, and rye, *Secale*) sprouts in 2–3 days; rinse 2–3 times daily and harvest when 2in/5cm long. Wheat sprouts in 4–5 days; rinse 2–3 times daily and harvest when 0.5in/1.2cm long.

Lentil

SEED

Lens esculanta. Use whole viable seed. Sprouting takes 3–4 days in the dark. Rinse twice daily. Harvest when 0.25–0.5in/0.5–1.2cm long.